D0881350

THE
CHIMPANZEE
WHISPERER

A LIFE OF LOVE AND LOSS,
COMPASSION AND CONSERVATION

STANY NYANDWI
WITH DAVID BLISSETT

Foreword by Dr. Jane Goodall, DBE

Arcade Publishing • New York

First Edition

Arcade Publishing books may be purchased in bulk at special discounts for sales promotion, corporate gifts, fund-raising, or educational purposes. Special editions can also be created to specifications. For details, contact the Special Sales Department, Arcade Publishing, 307 West 36th Street, 11th Floor, New York, NY 10018 or arcade@skyhorsepublishing.com.

Arcade Publishing® is a registered trademark of Skyhorse Publishing, Inc.®, a Delaware corporation.

Visit our website at www.arcadepub.com.

10 9 8 7 6 5 4 3 2 1

Library of Congress Cataloging-in-Publication Data is available on file.
Library of Congress Control Number: 2021945602

Cover design by Erin Seaward-Hiatt
Cover photo: Barbara Hollweg

ISBN: 978-1-950994-32-8
Ebook ISBN: 978-1-950994-42-7

Printed in the United States of America

For Nowera, your love and faith and strength have kept
us as one, despite hard times.

For Lou and Debby, who opened the doors between Stany and David.

And thanks be to God for the miraculous outworking of His plans.

CONTENTS

AUTHORS' NOTE

The International Union for Conservation of Nature (IUCN) has called on scientists, caregivers, and conservationists to stop publishing material that depicts humans in close contact with nonhuman primates. Stany and David wholeheartedly support these sentiments. Stany is a chimp specialist, and he only ever interacts with chimps in a manner that benefits them. Chimpanzees are dangerous, and Stany's behavior should not be copied or emulated in any way. Stany and David will never knowingly support any organization or individual that treats apes as pets or in a manner that is not in their best interests or true to their species.

FOREWORD

It was in 1960, more than a half century ago, that I began my study of chimpanzees in the forested mountains above the shores of Lake Tanganyika at the Gombe National Park. Even back then we knew a good deal about the similarities in behavior between humans and chimpanzees. But now, following long-term studies at Gombe and other field sites, we know a great deal more. We understand the strength of the enduring family bonds between mothers and offspring and between siblings that can last a life of more than sixty years. We know that chimpanzees, like ourselves, are capable of violence and brutality as well as love, compassion, and altruism. Through advances in science, we now know more about the biological similarities in the immune system, composition of blood, and the structure of the brain. We have learned that the DNA of humans and chimpanzees differs by just a little over 1 percent. The primates we call chimpanzees, *Pan troglodytes*, are the closest surviving relatives of the primates we call humans, *Homo sapiens*.

I loved working in the forests of Gombe, but I left to try to help efforts to conserve chimpanzees and their habitats. Their forests are being destroyed. They are hunted both for the live animal trade that captures the infants (by killing the mother) to sell to zoos or as pets, and for the "bushmeat trade," the *commercial* shooting of wild animals, including chimpanzees, for food. When mother apes are killed for this trade, their traumatized infants are often sold in the markets as pets. Then they need our help.

When people think of the champions of African wildlife, the names that spring to mind are usually Dian Fossey, George Schaller, David Attenborough, Iain Douglas-Hamilton, and others—Americans and Europeans. But there are an increasing number of true champions of wildlife among the African people as well. These include park wardens who resist rampant corruption and game rangers who risk and too often lose their lives in the fight against poaching.

This book tells the story of one of these inspiring protectors of African animals, Stany Nyandwi. From the time he first joined the staff of the Jane Goodall Institute (JGI), Stany has taken up the chimpanzees' cause. He has helped look after the orphaned chimpanzees in our care, and it almost cost him his life. He has endured untold hardships and personal sacrifice to care for these animals threatened by human wars, habitat destruction, and illegal hunting.

It all began for Stany in Bujumbura, the capital of Burundi. This is where I, too, first came face-to-face with the awful reality of the chimpanzee pet trade. The then American ambassador to Burundi, Dan Phillips, and his wife, Lucie, begged me to visit Bujumbura, where they told me there were a number of chimpanzee infants being kept as pets, often in terrible conditions. Before we could persuade their owners to release them, it was necessary to find them accommodation. The first two, Poco and Socrates, were put in a big cage built especially for them in the backyard of the embassy residence. Next, we managed to raise money to build a halfway house—a facility where we eventually kept twenty-two youngsters while searching for a location and the money for a proper sanctuary.

It was at this point that Stany came into the picture, when he joined JGI to help look after the growing chimpanzee family. Little did he know that this job would change the course of his life. It was clear from the start that he had a real gift for working with these creatures. He empathized with them and quickly came to understand the posture, gestures, and sounds that make up the chimpanzee "language." He was able to communicate with them in a special way to the point that we called him a "chimpanzee whisperer." Since that time, I have heard so many stories about his skill in resolving conflicts between individual chimpanzees—also stories about how he could calm tense or nervous individuals and reassure the infants who arrived traumatized and often wounded, having been torn from their dead or dying mothers. He was frequently able to help fellow caregivers when they had difficulty coping with problem chimpanzees.

This relationship between human and chimpanzee is fascinating. It becomes even more extraordinary when set against the backdrop of a brutal civil war that tore Stany's country apart, claiming the lives of his parents and siblings and separating him for four long years from his wife and children. He had been with us for five years when fighting erupted in Burundi between the two main ethnic groups, the Hutu and the Tutsi. As is always the case, hundreds of innocent people suffered. Two of our staff were killed; others, including Stany, were beaten. That he survived was a miracle. It was clear we had to close the sanctuary, which meant that somehow, we had to relocate the chimpanzees to a safe place.

At this time Debby Cox arrived to take over the running of the halfway house. Little did she know when she agreed to join us that her first task would be to help JGI organize planes to transport twenty chimpanzees to Kenya. Debby of course chose their best human friend, Stany, to go with them. He could reassure them after the gunshots and a frightening plane journey, and he became solely responsible for their daily care.

The Kenya Wildlife Service had agreed to house them while a sanctuary was constructed for them. My friend Russell Clark, a manager with Lonrho, the large conglomerate specializing in mining, agriculture, and hotels, agreed to build the chimpanzees a sanctuary on land where they operated Sweetwaters Serena Camp just outside Nanyuki. It is part of the Ol Pejeta Conservancy, which is still home to many endangered species including elephant, rhino, lion, buffalo, and leopard, as well as chimpanzees.

When their new home was ready at Sweetwaters, Stany went with them to settle the chimpanzees in and help with the training of the new caregivers. He stayed for three months. By that time a new sanctuary manager had arrived, and Debby had been appointed to oversee the development of another sanctuary in Uganda. She needed Stany to help her, so he began a new life working with the orphaned chimpanzees of what would become the Ngamba Island Chimpanzee Sanctuary.

I have been to that beautiful island on several occasions, renewing my friendships with the chimpanzees and the staff. Now there is accommodation for a few visitors in addition to the staff and spacious cages where the chimpanzees sleep at night in hammocks suspended high near the roof. On my last visit I got up early to watch Stany and his staff give the chimpanzees their breakfast of porridge and fruit before they were let out for their day of roaming the forest. Then there was time to sit with Stany to reminisce about the history of sanctuaries and some of the chimpanzees. I was able to share news of his old friends from Sweetwaters, Poco and Socrates and the rest.

Stany has since moved on from Ngamba Island, but he continues to work for JGI and for Africa's precious chimpanzees. Stany spent a year as a JGI Uganda education officer under our successful Roots & Shoots program. He worked with local villagers in Northwest Uganda, helping them coexist with nature in a way that benefits both them and the chimps. When our JGI South Africa sanctuary, Chimp Eden, required additional expertise to manage their community of orphaned chimps, Stany was the first person who came to my mind. Today, he continues to work with these special chimps, improving the quality of their lives and giving them the best possible alternative to a natural life in the forest.

I hope you will be fortunate enough to visit Ngamba Island or Chimp Eden someday. And I hope you will be inspired by this life story of a gentle, unassuming African man whose devotion and sacrifice have given new life to so many chimpanzees whose families, like his own, were torn apart by violence. If these chimpanzees could speak, they would acknowledge Stany as their hero, and more importantly, as their friend.

JANE GOODALL, PhD, DBE
FOUNDER, THE JANE GOODALL INSTITUTE &
UN MESSENGER OF PEACE
www.janegoodall.global
www.rootsandshoots.global

THE
CHIMPANZEE
WHISPERER

Prologue

A SPOT OF TEA AND AN ESCAPEE

It's a tepid, silvery morning on a small jungle-clad island, arrow-shaped and afloat in the liquid restlessness of Africa's largest lake. This island is home for first cousins, though regrettably, not a family at peace. For one has been the recipient of the worst kinds of atrocities the other could devise.

This is an island for refugees, a refuge for orphans. It is an orphanage for victims of our indifference.

There will be an incident today. Though not unprecedented, it will have a profound impact on all here who witness it—and on those who will hear the story retold in the years to come.

A whistle blasts, breaking the island tranquility, the calls of the weaverbird, and the washing of lake waves. The whistle means there's been an escape.

The escapee is Eddie, one of this island's victims of abuse and mistreatment. He still carries the scars . . . sometimes rather badly. He has a reputation as powerful and aggressive. He likes to display, running and screaming, with his hair standing on end to make himself look much larger. He likes to throw anything he can set his hands on. He likes to escape from his side of the island, and the electric fences are no real barrier. And Eddie knows the humans on the other side of the island are scared of him. How ironic, considering his cruel early years.

Strict protocols exist when there's an escape. The folks here know the danger. There was that incident in Connecticut when Travis came close to killing a woman. Before he was done, Travis removed the lady's hands and nose and eyes and lips. He shattered her facial bones and inflicted brain damage. Travis was supposed to be a pet.

But chimpanzees do not make good pets.

And now Eddie has escaped. He's no pet but a fully grown male chimpanzee, and he's just as dangerous. Everyone on the island moves down to the lakeshore. Chimps can't swim, so if Eddie comes too close, the humans can always wade out into deep water. Still, everyone is anxious. Some are afraid.

But there's always an exception. . .

An American man by the name of Jim is one of those standing by the lakeshore. He's here to see the chimpanzees of the island, but now he's afraid one might get a little too close. One of the island staff members stands beside Jim. There's nothing immediately remarkable about this man; he's maybe forty years old with hair shaved down to his cocoa-colored scalp; stocky in green cotton overalls and black rubber boots; a smile as wide and luminous as an African horizon. He produces a red thermos and a pair of white ceramic mugs, and he pours Jim a spot of tea. Jim feels slightly shocked, given the circumstances, but there's no fear to be found here. For this man is indeed remarkable, beyond any appearance. This man has a unique gift. This is the man they call the "Chimp Whisperer."

A vet arrives at the lakeside with a dart gun. If he can sedate Eddie, the chimp can be moved back to his side of the island. That's the plan. But the vet is afraid of Eddie. He refuses to go too close. He fires a dart. He fires a second. And both shots miss. Coming under fire makes Eddie furious. This situation is spiraling out of control.

And then, into this hurling, leaping maelstrom of black hair, teeth, screams, and testosterone strides the Chimp Whisperer. He puts down his tea and slowly walks up from the shoreline in those black rubber boots. He carries no gun, no club or stick. He is armed only with his head and his heart. And Eddie is waiting for him.

At one point, the Chimp Whisperer turns and finds that Eddie has been following him. The chimp is close—only about a hundred yards away—and clearly agitated. He is standing upright on his legs and swaying from side to side. His hair is erect. His mouth is drawn tightly in a severe grin. The Chimp Whisperer takes a deep breath. He considers his situation carefully. He knows Eddie. He knows himself. He knows what the score is. He suspects Eddie does too. The Chimp Whisperer gives a call, known as a pant-hoot. It's like lighting a fuse.

Eddie leaps forward, running on his feet and knuckles. More than 120 pounds of bristling hair and tensed muscle leap and run—straight at the Chimp Whisperer. Jim and the other people by the lake feel sure they are about to witness a tragedy.

No one would blame a man for running. To stand one's ground in the face of a chimp charge defies logic. But the Chimp Whisperer knows running is a bad idea. Running may only make things worse. So, he stands his ground. He stands his ground even as Eddie comes upon him.

The Chimp Whisperer has seen something the other humans haven't—maybe they can't—and instead of mauling, pummeling, and biting, the charging chimpanzee throws himself into the embrace of a friend. Eddie leaps into the Chimp Whisperer's arms. It's a maximum-impact hug. A category-five cuddle. The escapee has found a friend.

And down on the lakeshore, the people share looks of amazement.

"It's a moment I'll never forget," Jim says to me more than a decade later as we chat quietly and those memories return. "Yeah, he's the man. He really is the Chimp Whisperer."

So, who is this man who makes tea in the middle of a crisis?

Who is this man who can stare down and then make peace with a charging chimpanzee? Surely, he must be someone rather unique—someone special.

"I am Stany," the Chimp Whisperer will say when he introduces himself to you. And he will give you that luminous smile, and your day will suddenly feel just that little bit brighter. It's such a simple greeting . . . for such an extraordinary man.

For more than a year—from 2019 to 2020—Stany Nyandwi told me his stories. And he had so many stories to share—far more than one book could possibly contain. There was laughter and there were tears. More than anything else, there was love. Today, Stany and I speak as brothers. We're brothers half a world apart, brothers from another mother.

What follows are Stany's recollections from the past fifty years in his own words. His has been a life of incredible contrasts—of triumph and tragedy; of education and injustice; of heavenly highs and hellish lows. I hope his stories will inspire you as much as they have moved, challenged, and inspired me.

DAVID BLISSETT

Chapter 1

GROWING UP IN A LAND OF BROKEN HEARTS

I do remember that day on the island—the day Eddie escaped. Though it was more than ten years ago, I remember standing down on the shore of the lake with Jim and having a cup of tea before I went up to look for Eddie. I remember walking up from the lake, step-by-step, calling out to Eddie to try and win his trust. Once I did this, I knew I could lead him back to the correct side of the electric fence. And that's what I did. All was well! Yes, I remember that day.

Eddie escaped a lot. He was a big male chimp, high-ranking and strong. He knew about digging under electric fences to make an escape. He liked to throw things at people. On one of his other escapes, Eddie chased an Italian lady off the toilet before catching her hiding in the bushes. Then there was the time when he disappeared into the forest of the island for days and days. We searched everywhere for him. We worried that he'd been killed. But no, Eddie had been hiding away in the bushes—having a nice long honeymoon with one of the females. Eddie was a very naughty chimp. And Eddie was dangerous, but not to me . . . because Eddie was my friend.

So, how did I become friends with chimps like Eddie? How can it be that, after more than thirty years working closely with chimpanzees, I have never been seriously injured? And why did people start calling me the "Chimpanzee Whisperer"? To answer these questions, I need to go back in time more than fifty years, back to the land of my childhood, to a place many hundreds of miles away from that island sanctuary where Eddie still makes mischief today.

I was born in a heart-shaped land, in the center of a great continent. It's a land that looks a lot like paradise. Lush jungle mountains rise to the clouds and give birth to royal waters, for one source of the Nile is found here. In this green, rich land, they say almost three thousand species of native plants grow in soil that is fertile and the color of chocolate. And people can grow almost anything here—coffee and tea, bananas and mangoes, fields of sugarcane and basket loads of fresh vegetables. Truly, this is a wild and beautiful land. It might be Eden.

However . . .

I was born in a heartbroken land, in the center of the great continent. Beyond the natural beauty and earthly riches, there is something much darker. In this country, there is immense sadness. There are the tears of injustice and the fire, blood, and bone of warfare and hate. In this land lives something very dark, something born in the hearts of human beings.

Truly, these things are both representations of the land of my birth.

Burundi is a landlocked country in the central east of tropical Africa. When you look at a map of Africa and you find Burundi, you can see that it's shaped like a human heart. Someone told me Burundi was only about the same size as the state of Massachusetts in the United States. I suppose that's not very big for a whole country, but it's big enough.

On Burundi's borders are the Democratic Republic of the Congo, Rwanda, and Tanzania. The capital city is called Gitega and is in the center of the country. It wasn't always. For many years the capital of

Burundi was a city called Bujumbura, which is on the shores of Lake Tanganyika in the northeast. And while it's not the capital anymore, Bujumbura is still the largest city in Burundi. It's a city that I came to know well.

With rich volcanic soils and a tropical climate, Burundi should be one of Africa's most prosperous lands. However, like many other countries in Africa, Burundi has been ravaged by many years of tribal hatred and civil war. There has been horrific genocide. And even today, almost sixty years after independence, Burundi is still a divided and troubled land.

I was born in March 1968. I believe a lot of things happened in 1968—there was the Cold War between the West and the Communist countries. America was fighting in Vietnam, and Communist armies had invaded Czechoslovakia. The Summer Olympic Games were held in 1968 in Mexico City. Back then, Burundi was still a young nation, moving towards an uncertain future.

My future began in a rural village in mountains about thirty miles north of Bujumbura, which was the capital city. My parents were named Simon and Silvia. They were married in the 1950s, though I don't know the exact year. That was before Burundi became an independent country, back when my homeland was a colony of Belgium, a country in faraway Europe.

My parents' first child was a boy. This was very good for them, because in Burundi at that time, boy babies were preferred. In our culture, boys grew up and inherited land and other family property, but girls usually left the household when they married, and they joined with the family of their husbands. The joy of my parents' first child did not last for long. The little boy died when he was still only an infant.

Simon and Silvia did have other children. They had many children—a lot of them girls. This was a mixed blessing, particularly for my mother. Failure to have a boy child could be viewed as weakening the family. And weakening the family could threaten a woman's future. It could be grounds for divorce. Burundi was poor, and divorced women would face terrible hardships, especially if they had children to care for. Faced with this dilemma, my mother turned to God.

Every day, as she worked in the family garden, raising her hoe and striking the blade into the earth, my mother would pray for a boy. Then, she would get pregnant and have a baby . . . and it would be another girl. She had six baby girls in a row! With each birth, I think my mother became more and more worried about her future.

So, in a very real way, the birth of my mother's seventh child was a big blessing. I was my mother's seventh baby. I was always close to my mother. She always cherished me, and I think that's because my birth may have saved her life. I was the boy baby she'd been praying for.

In my family and culture, people usually have more than one name. However, we don't have family names like you might have in America or Australia or the United Kingdom. I was called Stanislaw, which is a European name that was quite popular at the time. This was shortened to Stany. My other name, Nyandwi, is a word in the Burundian language—Kirundi—and it means "born seventh." The sister born before me was named Miburo, which means "Nyandwi is coming."

Life really changed for my mother after I was born. Her next two babies were also boys. My parents named the next boy Minani—meaning "born eighth"—and then they had another boy, who they called Nyabenda, which means "born ninth."

So, six girls and three boys . . . yes, I grew up in a very big family.

To begin to understand what happened in Burundi—and in neighboring countries like Rwanda—you first must understand the cultural and tribal environment. Even today, Burundi has a population mostly made up of two cultural groups—the Hutu and Tutsi.

The Hutu are the largest cultural group in Burundi, and this is the group to which I belong. Hutu people tend to be short, thickset, with broad noses and dark chocolate-colored skin. Known for our agriculture, it is thought the Hutu came from the west, moving across central Africa and settling the land as they went.

Tutsis are taller people, lighter colored and with thinner bodies. It is thought that the Tutsi migrated to central Africa from the east—from places like Ethiopia. They are known as cattle herders and as brave, strong warriors.

Even though the tribal groups rarely interacted with each other, there was no real conflict. The land was relatively peaceful until 1897, when the Germans arrived and colonized the region. Germany considered Rwanda and Burundi one territory, which they called Ruanda-Urundi, and they governed at a distance from Europe. The Germans mostly favored the Tutsis over the Hutu. This sowed seeds of tension and unrest. Those seeds would grow to become the thorns and weeds that continue to choke my country, even to this day.

After World War I, Germany lost control of Ruanda-Urundi, and Belgium was given the right to rule. The Belgians took up where the Germans left off. They employed a Tutsi aristocracy to govern the territory, and this caused even more tension and disputes between Tutsis and Hutus.

Then, in 1962, six years before I was born, Burundi and Rwanda broke away from Europe, and the people voted to become separate, independent countries. In the early years of Burundi, many good people tried to create a balanced government, with representatives from both the Tutsi and the Hutu. This peaceful coexistence did not last for long.

When I was a boy—and it is still true today—the Hutu were the largest cultural group in Burundi, but the Tutsi were the people who mostly had the power. Tutsi leaders controlled the military and the government for many years. Sometimes, these leaders used their power badly, imposing hardships on Hutu people. Conflict between Tutsi and Hutu became an everyday part of life in Burundi. The suspicion and hatred cost my country a chance at real prosperity. There would be terrible bloodshed. Hundreds of thousands of people would die.

Time after time, the hatred between Tutsi and Hutu would escalate. There would be isolated instances of violence, or battles in a small part of the country. But sometimes, the hatred erupted into full-blown civil war. In my lifetime, there have been several large civil wars and mass genocide. And, you know, very few people outside my country even knew those things were happening.

I knew exactly what was happening. Burundi's nightmare was part of my reality. Someone once said that civil war bedevils Africa. I believe this is true. When there is violence and hatred, no one wins. Burundi is proof of this.

In 1972, the first major civil war erupted across Burundi. Even though I was only four years old, I remember running and hiding with my family. I remember seeing soldiers shooting people from a helicopter.

People were running. The helicopter and the guns made a terrible noise. People were screaming. I was only a boy, but I can still remember it.

I'm much older now—that was almost fifty years ago—but the cost of so much war has been a heavy one for many families in Burundi. My own family has suffered great loss. Even today, almost sixty years after independence, Burundi is still a land without true peace. It is still a land of broken hearts.

My childhood years might have been complicated for the adults of Burundi, but for most of the time, the life of a Hutu boy was simple. My family lived in a two-room hut made of mud and straw, with straw used for the roofing. Only the poorest of poor families lived in these straw huts, and we were very poor. We had no electricity and a single communal well for water. We had only a few basic clothes. I was fifteen years old when I wore my first shoes—and they weren't even a matching pair.

The nearest medical facility was a Catholic mission clinic. This clinic was a three-hour walk from our home along a rough dirt track. The only way to get to the clinic was by foot. The sick and the injured had to walk. And if a patient couldn't walk, then they had to be carried on a stretcher. This was no hospital. The clinic was operated by nursing nuns who could provide only basic care. There were no doctors. So, if you were badly injured or you had a serious illness, there was a high chance you would die—whether you made it to the clinic or not.

I know that for many people in the world, the 1960s and '70s were years of great change. I know that many modern things were invented, like spacecraft and computers. But for us, in our village, the only time we saw something modern was when the local parish priest came to visit. The priest would sometimes come to the village to hold services. The priest owned a car. Oh, that car was really something. We would walk up to it carefully. It was all shiny and amazing, and we just wanted to touch it. We certainly weren't allowed to climb inside, and it wasn't until I was thirteen years old that I had the chance to ride in a motorcar.

While we did have neighbors living close by, ours wasn't the type of village you might imagine. There were no stores or marketplaces and no real community facilities. Different families lived in their own small huts, on adjacent plots of land where their crops were grown. There were groves of banana palms, sugarcane, and fruit trees. I can remember warm and humid days, when the air was thick with smoke coming from all the cooking fires. My mom and sisters would cook our food over a wood charcoal stove. The communal well was near our hut, and the women would draw water up from the well and carry it back to the hut in clay pots balanced on their heads. My family had a small plot of land, where we grew vegetables and bananas. We also grew coffee. Everyone grew coffee, because the government made it compulsory. And they say Burundian coffee is among the best in the world. Growing coffee should have made my family rich. It didn't.

To help our family survive, my father took on a paid job. This was good in one way, but very bad in another. Paid work was a rare and valuable thing in Burundi at that time. Most Hutu families lived off their land, and in poor times they would struggle to even feed themselves. So, my father being able to earn money was a bonus.

My father took a job as a police officer. He became a Hutu man enforcing Tutsi law. That was a bad thing. I know this job cost my father respect among other Hutus. One of the most unpopular parts of his job was collecting a census tax, which cost people three thousand schillings (around $6 US) each year. If people could not pay the tax, a police officer might accept a smaller payment as a bribe. Anyone refusing to pay the tax—or a bribe—risked being beaten or tortured. Tax collecting meant police officers were often hated by the people, especially Hutus. I think this was stressful for my father. It was certainly stressful for us.

Drinking in Burundi is common. My father was a heavy drinker. He would get really drunk. And when he was drunk, he became violent. The stress and anger from his job were taken out on his family, including the children. I remember being beaten. It was scary. Family violence became part of life when I was a boy.

While my father was away working, my mother and my sisters, my brothers, and I would tend our garden and keep house. Life was casual, and there was no list of chores to be done. Sometimes, my mother would need something, and she would say to me, "Hey you . . . go and get some firewood." Most days, though, I didn't have any real chores to do.

I loved going out into the countryside. We had an endless playground on our doorstep. I was a little bush boy and loved playing in the forest with my friends. We climbed trees, and I used to shoot all kinds of animals with a slingshot I'd made. If I hit a bird, like a pigeon, maybe I would take it home for my family to cook, but mostly it was just for fun. Later in my life this behavior became very interesting to me—this idea of killing for fun. Like bush boys, young chimpanzees—almost always males—go wandering in the forest and kill smaller animals. This type of hunting is mostly not for food; they do it just for fun. And we were like that too.

There were all kinds of animals in the bush back in those days, such as monkeys and bats and birds. But today, in Burundi, most of those animals are gone. And almost all the forest has been cleared away.

When I turned ten years old, I had the opportunity to go to school. It was a six-mile walk to the nearest school, and the journey took two hours. We'd walk along a narrow path up through the mountains, past villages and rolling fields of grass or groves of crops, and through small patches of forest trees. We would walk in groups, happy kids in our shirts and shorts, dresses or skirts, most of us barefoot and all filled with noise and youthful energy. And then, at the end of each school day, we would all walk home again. It would only take us about one hour and a half to get home. We would run. It was easy. It was downhill all the way!

Ours was a mission school, partly funded by the Catholic church and partly by the government. The building was simple. It had a dirt floor. There were chairs, but no desks. And there were certainly no schoolbooks. At the front of the classroom there was a blackboard on which our teacher would write words. Each student had to bring a small wooden board and a stick of charcoal for copying these words. The goal wasn't to give us a full education, but to teach us to read and write in the official language—Kirundi.

School days went from eight in the morning until four in the afternoon, but not all that time was devoted to lessons. When there were breaks in class, we had to work. We went outside and started to dig up stones. We carried the stones over to a place where a new school building was being constructed. We'd spend up to two hours each day carrying stones. It was hard work. I guess it was really slave labor. But I was only a child, and I thought it was a normal thing to do. I didn't know school could be any different.

There was no cafeteria at our school either. There were no lunches. If you didn't bring something from home, like a piece of sweet potato, a banana, or some sugarcane, then you went without food for the whole day.

When it came to play, we didn't have any toys. If kids wanted toys in our village, we had to make our own. I used to make balls out of banana fiber. I would roll leaves up and then bind them together to form a ball. I would make a ball in the evening, and then I would take it to school next morning. When I took a ball to school, you know, I was special. I could select which friends I wanted to play with. We would play football (soccer) out in the field. But the ball wouldn't last very long—maybe only a few hours being kicked around—and then I would have to make a new one. And then, something special happened.

One day I went to school with a tennis ball. I had a little money and I bought this ball from a boy who had come back to our village from Bujumbura. I took the tennis ball to school and that day, I was like a king. Nobody in the school had ever seen a real tennis ball up close. The other kids were so excited. I mean, some of them just wanted to come up and touch it. We did not have any proper balls at school to play with, so this was all new to us. That day, I was really popular.

I was okay at school. I learned to read and write in Kirundi. Most important of all, I was taught how to learn—it wasn't just about copying down words. I think this served me well in later life, because today I can speak five languages: Kirundi, English, Swahili, and native languages from Uganda and Rwanda.

As it happened, I didn't stay long in school—only three years. I was thirteen when I stopped going. My father didn't respect education. He believed that Hutus with education were more likely to be killed. And I found out later how right he was.

When I was much older, I heard a story about twenty kids from my local area who were going to secondary school. These were Hutu kids. They went back to school after the holidays, but their school was far away. You know, this was about 1993, and everywhere you went at that time was dangerous. The civil war had come. There was fighting and killing. Still, the schools were open, so the children went.

Those twenty kids got on board a VW minibus to go to school, but that was the end for them. They never made it. Gunmen, probably Tutsi soldiers, stopped the bus, and the kids were massacred. Innocent Hutu children, killed in cold blood, just because they were going to school. So, I think now that my father was right. Because I didn't stay long in school, I had the chance to grow up.

The fun and simple days of childhood—playing with soccer balls and slingshots—don't last forever. As I moved into my teenage years, my life was about to change. I needed to go to work. Life would never be the same.

Child labor was—and still is—common in many parts of rural Africa. In our village, young people were expected to work in the family garden or to search for paid jobs so they could earn money. When I was about thirteen, around the same time as I left school, I joined together with some other kids, and we began to sell sugarcane.

We would go to someone who grew cane and we would buy it from them. Then we would carry the cane to market. This was a hard job. It was a long way to market, and fresh sugarcane is heavy. Each of us could carry only four or five canes. I enjoyed being with the other boys from the village, but we didn't make much money.

The job took two days. One day we bought the cane and took it home. Then we would get up early in the morning and carry the cane to market. It was about nineteen miles to the nearest market and we walked the whole way. Many times, we would leave home with nothing to eat. If we became too hungry we could always cut a piece of sugarcane on the way and eat that, but we had to be careful not to get greedy—we didn't want to eat too much of our profits! Sometimes we would earn only enough money to buy something small, like some bread. Whatever money was left, we took home to our parents. They would be upset when no money came home, but when we brought back something there was much joy.

Sugarcane wasn't the only thing we would sell. If we couldn't buy any cane, we bought and sold bananas. We would buy bunches from other families or cut them from our own trees. My sister and I would carry them with the other children all the way into Bujumbura to sell.

These journeys were an adventure for us. We'd carry tied bundles of cane, or big bunches of bananas, down by our sides or up on our heads. Once again, the long treks to market were made entirely on foot, along rough bush tracks, and we didn't wear any shoes. And there are some really deadly snakes living in Burundi—cobras and mambas and vipers! We would get up at three or four o'clock in the morning, and it would still be dark when we set off. The path was dirt, and if it rained it would become slippery. If you fell, you could easily break some bones. And while it was a long way to walk, we were all used to it. It wasn't so bad.

The city could be dangerous as well. I remember the first time I arrived in Bujumbura with my sister. It was really exciting, but even though I was thirteen years old, I had to hold my sister by the hand while we crossed the street. She didn't want me to get run over. We would sell our bananas at the market, and then we would walk back home again.

And then, on one visit, my sister and I did something really exciting. My sister told me she didn't want to walk all the way home. She wanted to catch a taxi. This sounded amazing to me, but we didn't have much money, so we could only afford to go a short distance—I guess it was a couple of miles. The taxi was a VW mini-bus. I was thirteen, and it was my first time riding in a car. Oh, I got dizzy, because when I looked out the window it felt like the trees were rushing backwards. I held on to the seat real tight. After we got home, I had such a wonderful story to tell, but I can't remember if we got in trouble for using all our money on a taxi ride. We probably did!

My time selling food at the markets didn't last long. I soon discovered there were other ways to earn money for myself and my family, and these didn't involve long treks to and from the city.

I was fifteen years old, and I remember one of the boys from our village coming back after being in Bujumbura. Oh, you should have seen him. He looked so different. He had new clothes and a wristwatch and a great pair of shoes. We all wanted to be like him. He told us he was working as a houseboy in the city. You know, when I think about it today, I suppose he was sent by the people he worked for to bring in new boys. But at the time, all I could see was that beautiful watch.

When this boy told me I could get a job like his, I was really excited. And my mom and dad were happy for me to go, because we were so poor. They felt this job would be a way for me to earn some extra money for the family.

Now, this was in the eighties, when Burundi's Hutu population was living under the authority of the Tutsis, and the education of Hutu children was not encouraged. With not much schooling, many young Hutus left their villages to work in cities like Bujumbura. What I didn't know was that these jobs meant hard work and little pay. This was going to be the start of a difficult time.

I left home and went with the boy to see a Tutsi family in Bujumbura. This family gave me a job as a houseboy. It didn't take long to find out that "houseboy" was the wrong name for this job. I had to do whatever the family wanted me to do. I would wash clothes and dishes and clean dirty baby diapers, plus I would clean the house and the compound. That was okay, I was young and fit and I could do the work. But I was not treated like a boy of the house. I wasn't considered a member of the family. I was more like a possession. I was a slave.

Sometimes the family wouldn't even pay me for my work. I was fed table scraps. There was no bed, and I had to sleep on a woven mat on the floor. I wasn't allowed to do anything with the family, not even the children. When I wanted to go home and see my own family, they refused to let me. But, you know, I was still only fifteen. I just did what they told me. It was only later that I understood how badly they had treated me.

Eventually, after they kept refusing to pay me for my work, I decided to leave this household. It was dangerous to leave, and if I'd been caught I would have been severely punished. So, I woke early one morning, before dawn, and I snuck away into the city. Then, instead of going back to my village, I went looking for another job. I knew my family still needed the money. I found myself another houseboy job, with another Tutsi family. But this second job was just as bad as the first. I suffered the same neglect and abuse.

I soon became a young man going around and around in endless circles. I was caught in a cycle of slavery with different Tutsi families. I worked for many families. I don't really remember how many there were, maybe at least ten. Sometimes I only stayed a few months, and sometimes I stayed for longer. This went on for four years, until I was nineteen.

Although these first jobs were hard, and although they were more like slavery than true paid work, they did provide me with some income. More importantly, working as a houseboy led me on a path to somewhere better. I didn't know it at the time, but I was about to open a doorway to a life I could never have imagined. I was about to discover a gift I never knew I possessed. And I would face dangers no one would believe possible to survive.

I was about to take a journey to the halfway house.

Chapter 2

HOUSEBOY FINDS HALFWAY HOUSE

It was the 1980s. I was still a nineteen-year-old Hutu man working for Tutsi masters. Many of the houses I worked for didn't have water piped inside. This meant you had to go and collect water from a public tap. I remember waiting in line to use this faucet. Sometimes bigger boys would push in front of me, and I would have to wait even longer. Waiting too long was bad, because bringing water back to the house late, even if it wasn't my fault, could mean punishment. I filled metal cans with water and then carried them to the house. Those cans were heavy, and I carried so many over the years, I tell people today that's why I'm so short!

Each time things became too bad at one house I would sneak away and find another. Eventually, I found myself at the house of a wealthy man who also owned hotels and restaurants. This man wanted a house-boy and a cook. I had learned how to cook as a houseboy, so I went to this man and I got a job with him. But, though he hired me for two jobs—cook and houseboy—I certainly wasn't paid double wages.

One night, the boss man's car broke down in a bad part of Bujumbura, where robbers and bandits used to travel on their way into the city center. Because it was in such a dangerous place, the boss man's driver refused to stay in the car after dark. I was ordered to go out and sit in the car overnight to stop it from being stolen. It was a long night.

I knew that a car would be a good thing for robbers to find. I also knew they wouldn't care if someone like me was in the way. I thought people like that were going to come and steal the car and I was going to die out there in the dark. I was terrified. But the boss told me to go, so I had to go. He also told me that I wasn't allowed to fall asleep or I would get into big trouble. You know, it was amazing, but all through that night, as I sat there and counted the minutes, not one person came up to the car. No one even walked past. It was a miracle! And I was relieved to see the sun come up the next day.

As well as his own home, this latest boss also owned another house that he rented to missionaries visiting Burundi from overseas. I made friends with an American missionary and his son, who happened to be staying at the property. And I also made friends with another member of the household staff during that time.

In Bujumbura, the people with money sometimes employed watchmen to guard their homes. Elie Nkurikiye was a Hutu man working as a watchman at the same house where I was houseboy. Sometimes I would hide food to give to Elie, because the boss man's family would not give him anything to eat, and he was very hungry. This is how we became friends.

I didn't know it then, but these friendships would lead me in a new direction. My life would change forever. It was through Elie, the American missionary, and his son that I first heard about Jane Goodall and her work with chimpanzees. After Elie stopped working as a watchman, he went to work at a new Jane Goodall Institute (JGI) sanctuary in Bujumbura. One day he came back to the house to visit. He told me about the sanctuary and what they were doing. I was fascinated.

*

There were small populations of wild chimpanzees living in the for-ests of northern Burundi, but I had never seen a chimp in the wild. And yet, I did know a little bit about them. There was a white man who had a compound near one of the paths I used to walk along from my village to Bujumbura. This man kept baby chimps in his compound, and I would see them when we went past. I also used to see chimps in the city itself. There were men who would carry little chimps around, and those chimps would do things to make us laugh. I used to enjoy watching chimps. To me, they seemed like funny little hairy people. What I didn't know was how those chimps came to be in the city. I didn't know about the cruelty that was happening. Not back then.

It was only after Elie went to JGI that I heard about what was really happening to chimps. Elie told me stories about an amazing English lady who worked with chimpanzees. He called this lady "Doctor Jane."

Eight years before I was even born, a young Jane Goodall arrived at the Gombe Stream Reserve in Tanzania. She had come to Gombe to study chimpanzees. At the time Jane Goodall began her work, people didn't know much about wild chimps. Even though she had no formal training, and no qualifications, Jane was to discover some amazing things about these human-like creatures. Jane found out that chimps have complex social and family lives. She learned that chimps show human-like emotions such as revenge and loyalty and kindness. They are skillful hunters, and sometimes they kill and eat other ani-mals. Perhaps most amazing of all, her research showed that chimps are capable of making and using simple tools. Before Jane Goodall, everyone thought that only humans were toolmakers.

As Jane learned more about chimpanzees, she also began to see that they were disappearing from the wild. Forests were being cleared for agriculture and mining. In some places, they were being hunted for bushmeat, and baby chimps were being stolen from their dead moth-ers to be sold as pets around the world. The situation was serious. Jane began to spend less and less time at Gombe and more time traveling the world to campaign for the chimps. Then, in 1977, she started the Jane Goodall Institute (JGI) to try and save the remaining wild chimpanzees and their forest homes.

In the late eighties, Doctor Jane was invited by the American ambassador to go to Burundi and provide some conservation advice. Wild chimps were vanishing from my country. When she arrived, Doctor Jane was horrified to find young chimps being kept as pets in the cities, often by wealthy foreigners. Many of these chimps had come from Congo and had been smuggled out of the country when they were babies. These were the same chimps I had seen on my first visits to Bujumbura. Many of these poor babies were kept in very bad conditions. They were fed the wrong kind of food and were suffering without the company of other chimps.

After her visit, Doctor Jane started JGI Burundi, and this organization established a sanctuary that they called the "halfway house," not far from the center of Bujumbura city. The halfway house was where Elie went to work after he finished as my boss's watchman.

"Halfway house" was a very good name for the JGI sanctuary in Burundi. The chimps living at the sanctuary couldn't be returned to the wild. Because they had left the forest as babies, and because they were habituated to humans, these chimps didn't have essential forest skills. If they were released back into the forest, they wouldn't know how to find enough food to survive. They might be attacked by other chimps and chased out of the forest. Scared, starving, and without a natural fear of humans, they might have raided villages and become dangerous to people.

So, once they were surrendered or confiscated, these chimps were taken to the halfway house. And then, even if they couldn't go back to the forest, the JGI plan was that one day they would be moved from this place to a larger permanent forest sanctuary. So, you could say the house really was halfway to somewhere better.

Also, it wasn't just the role of the halfway house sanctuary that gave it the name. This wasn't a large area of grassland or forest for orphaned chimps to live. It really was just a house with a big backyard. The house was in the suburbs of Bujumbura, close to the international airport. In the backyard there was a large steel enclosure which was divided into sections by sliding doors and raceways. In this enclosure lived a group of larger adult chimps, whose strength and troubled pasts made them potentially dangerous to people.

There were younger chimps as well. They were free to roam and play in the backyard, climbing trees and swinging on play equipment. However, the mischievous youngsters had to wear collars attached to long rope leashes to stop them escaping the compound. The yard of the house was only fenced by thin bamboo stakes, and without the collars and leashes, the smart and nimble little chimps would be able to jump out of the backyard easily. And once they escaped, they would be exposed to the dangers of the city.

Sometimes, even with the collars and leashes, they would escape. There was one young female chimp at the halfway house called Dufatanye. She was a smart little girl, and she worked out ways to escape from the backyard. Once free, she used to break into the offices next door, scare the people away, and then drink their coffee. Oh, Dufatanye really loved coffee! But Dufatanye also had a habit of biting people. So, when she escaped, the caregivers used to have to quickly run next door and catch her before she started searching for the coffee and causing a problem.

I used to wonder how chimps like Dufatanye came to be at the halfway house. I thought that maybe JGI paid money to buy them back from their owners or the smugglers. But no, the chimps had either been confiscated by the government or surrendered to JGI. Doctor Jane has always said that JGI will never pay money to buy orphan chimps, no matter the circumstances. Paying for chimps would only encourage more poaching in the wild.

Even though the halfway house was far from perfect, the facilities were the best that JGI could manage. Most importantly, the chimps were now safe. They were being given good food, medical attention, and proper care. And one day, they would move on from the halfway house to a better sanctuary. Then they could really make a fresh start.

Tired of poor pay and mistreatment, I was ready to make a fresh start too. Knowing this, Elie asked the people at JGI if I could have a job. The boss was a man called Dean Anderson, who agreed to meet me. Dean was an American, and I remember how tall he was. Maybe he was the tallest man I'd ever seen up until that point. Dean and his wife, Susanne, who was a veterinarian from Australia, were both very kind people. After four years of cruelty, it felt wonderful to be treated so well.

I'll always remember my meeting with Dean. I couldn't speak much English, and he couldn't speak Kirundi or Swahili, but we somehow managed to communicate with each other. I think that maybe Dean just knew how to pick good people. I hope so, because he gave me a job at the halfway house.

Once I started working for JGI, I began to learn English. I learned from Dean and Susanne, but I also had plenty of opportunity to practice with the many volunteers who visited the sanctuary. There were volunteers from America and Australia, and they all spoke only English, so to do my job properly I had to learn fast!

I was employed as a gardener and groundskeeper at the halfway house. It was my job to clean the backyard and keep everything neat and tidy. When it came to the sanctuary's most important residents, however, that was another matter. I wasn't employed to work with the chimps. In fact, I was told to stay right away from them.

Workdays at the sanctuary started at 7 a.m., and we finished at 5 p.m. There was really nowhere to stay overnight at the house, so each evening I would make the thirty-mile journey back to my village in the mountains. The journey would take many hours. I often used to jog to get there more quickly.

If I could avoid walking—or jogging—sometimes I would hitch a ride. Elie had a bicycle, and I would ride with him until we reached his village, and then I would walk the rest of the way to mine. In the mornings, I would walk to Elie's village, and we would then ride down to the sanctuary together. But if Elie had a problem, or he was not going to work that day, then I would have to do the whole journey on foot.

At first, there were only a handful of confiscated chimps at the halfway house, but this would soon change. I remember more and more confiscated chimps arriving at the house. As the days and weeks passed, the staff and volunteers had to look after more and more chimps. Though I still thought they were cute to watch, I knew now that this was a sad situation. Those little chimps were evidence of the horrors that were occurring in the forests not so far away.

Those days were really crazy. The backyard of the sanctuary would be filled with young chimps, with their pale faces, big ears, mischievous brown eyes, and those cute little white tail tufts. They would be climbing and swinging and tumbling and spinning and running. They seemed to have endless energy. The caregivers would be there too, out of breath, being jumped or climbed on. They would be chasing the chimps, ready to intervene whenever their antics got them into trouble. Meanwhile, behind the red-painted bars of their enclosure, the larger chimps would go about their daily routines, eating and displaying, calling and climbing and swinging around.

And there I was, watching the chimps and the caregivers do their thing, while I went about my daily routine. I would wash and clean. I would sweep up discarded food, old leaves, and dirty straw. I would pick up the chimp shit. And then, gradually, something began to happen. Slowly, the distance between me and the chimps started to get smaller.

Chimpanzees share 98.7 percent of their DNA with humans, and if you spend time with them, you can see that. Working at JGI was the first time I was really able to watch chimps closely and under-stand how much like humans they really were. Yes, I had seen chimps before—in the yard of that white man's home, and on the streets of Bujumbura—but I had never seen them so close or for so long.

But soon there was a problem—the barrier between me and the chimps was starting to be tested, not by me, but by the chimps them-selves. I would go past them with a wheelbarrow full of rubbish, and I would see them watching me . . . watching . . . watching . . . and then, when I was close enough, one or two of them would reach out and try to touch me. Or they would throw their shit at me. I had never seen anything like that before—animals that threw their own shit. Remember, at this point, I wasn't even allowed to touch the chimps, but they didn't know that. And some of them very much wanted to touch me. I could see that they wanted to make friends with me.

You know, when I saw that happening—when I saw chimpanzees reaching out to try and make friends with me—it felt like something had changed inside. I can't really explain it. From that time on, I started to be really fascinated by them.

It would be six months before I was allowed to make contact with any of the halfway house chimps. I began to ask questions of Dean and Susanne or the other people working with the chimps. I took the time to watch and learn. I started to understand that each chimp was an individual, in just the same way each human is. And there was one youngster who made an extra effort to reach out and make a connection. He would become the first of many special chimpanzees in my life. His name was Max.

Max was a very nice boy. He'd been confiscated from a French film crew in 1990, when he was only two years old. For some reason, he was always attracted towards men, especially if they were in uni-form. Max is much older now and living in another sanctuary, but I believe that, even today, he still likes men in uniform.

Back then though, Max was young. That meant he was allowed out into the backyard of the house wearing a collar attached to a long rope. He could climb the trees and run around. I could see that Max really wanted to get to know me, but I wasn't allowed to touch any of the chimps. I would be walking along with my wheelbarrow, and Max would get all excited and try to run after me. Eventually, he worked out he could get my attention by throwing things at me. He would throw his food bowl or maybe a stick or a handful of shit. Yes, you could say that Max was a mischievous boy!

Finally, his persistence paid off. And I can honestly say, it is through Max that my friendship with chimps began. Dean Anderson agreed that Max and I could meet. I will never forget that moment. Max reached out and touched me and after he touched me, he began to smell his finger. Dean told me this was one way that chimps get to know each other. And Max kept doing it—he would reach out to touch me, and then he would smell my scent on his finger. To me, well, this felt so special. I think maybe it was special for Max too. Friendship for chimps is very important, just as it is for human beings.

After that day, Dean and Susanne began to teach me more and more about the chimps at the sanctuary. I was like a sponge. I wanted to learn as much as I could. Along with Dean and Susanne, there were many educated people volunteering at the house. There were university students who were studying chimps and who had volunteered to work for JGI so they could learn more about them. Now, I had done only three years at school and had never been to university, but I was still young. This was good, because it meant I could learn fast. It seems to me that young people often learn faster, and I know that I certainly did. I learned so much about chimps from listening and watching and asking questions. And as I learned, I found out there was so much more to know. Even today, after so many years, I am still learning about them.

I had heard people say that chimps bear grudges, like humans do. I think that's true in some cases, but it sure wasn't for one chimps at the halfway house. After Max, Poco was one of the first chimps I became friends with at the sanctuary. He had many reasons to hate people. For nine years he lived in a tiny metal cage outside a business in Bujumbura. Poco was being used by the owner to attract customers to his business. And, because he'd spent such a long time confined in such a tiny cage, after he was rescued and released, Poco was only able to sit down or stand on his two hind legs. This was quite an unnatural way for a chimp to walk, because most of the time they walk on all fours: on the knuckles of their hands and the soles of their feet.

Poco was one of the first chimps cared for by JGI Burundi. After he came to live at the halfway house, Poco became the alpha male—the boss chimp. And he was the boss when I first came to the sanctuary. Poco had a face like a loveable uncle, and he was always so outgoing and nice. I remember how much he used to like meeting new people. He would get excited whenever someone new came to the sanctuary. There was one volunteer with whom Poco was really in love. She was a young Australian woman, and Poco would get very excited whenever she went near him. His hair would stand on end, and he would get a huge erection. If that young woman went out in the sanctuary car, Poco would sit in his enclosure at a place where he could see down the driveway. He would wait there, watching out for her. And when she came back . . . oh . . . he would be so happy. He would keep jumping up and down and pant-hooting until she went over to the enclosure to greet him. One time he became so excited when she went up to talk with him that he had a little accident. So, yes, I think you could say that Australian woman was Poco's girlfriend!

One of the volunteers I met at the halfway house was a former Australian vet nurse and zookeeper named Debby Cox. After working for many years with chimps at Taronga Zoo in Sydney, Debby had devoted her life to saving them. Debby was, and is, a remarkable soul. Now, Debby is not a tall woman—she's even shorter than me—but she's as fierce as a lioness and just as determined. She's also very kind and very generous. I came to learn that Debby would do almost anything to save chimps. And, in 2009, she received one of Australia's highest honors—being named Member of the Order of Australia— for her services to chimps and conservation.[1]

I was to learn more about chimps from Debby Cox than anyone else. She knew so much. I would watch Debby, and I would listen and learn. Debby also taught me about serving others. She would never take any time off. Debby always did the same jobs as the African workers. She would clean cages and wash floors. She would even wash the staff's dirty dishes after lunch. Debby taught me how important it is to truly respect other people.

And, perhaps more than anyone else, Debby saw my potential as a chimp caregiver. Years later she said to me, "Stany, it was pretty clear that you had a special gift. I've seen you work with chimps you've never met before, and instantly there's a connection. There's trust and respect. That's very unusual. For most of us, that level of connection takes weeks, but not for you. You're very tolerant and patient. But there's more. You have something special. The chimps understand you, and you understand them."

It meant so much to hear someone like Debby Cox say that. Debby is my sister.

But back at the halfway house, before I even knew that I had any special gifts, I was about to be seriously tested. For the first time ever, my life would come to depend on talents I didn't fully understand myself.

No one remembers the exact date, but everyone remembers the incident. I was now working as a chimp caregiver at the halfway house. On this day, I was cleaning the large enclosure where the adult chimps lived. I went into the raceways and began moving chimps around the enclosure so I could clean. Larger chimps—especially the adult males—are too big, strong, and dangerous for caregivers to go

---

1 honours.pmc.gov.au/honours/awards/1139698.

into an enclosure with them. So, at the halfway house, we used to move the chimps around different parts of the enclosure to allow us to clean in the vacant sections.

Now, on this particular day . . . well, I really don't know how it happened. Perhaps I wasn't paying attention, but I somehow found myself locked in the same part of the enclosure as one of the bigger chimps—a male by the name of Safari.

Safari had spent the first years of his life living in a small outdoor cage at a hotel, where tourists would taunt and tease him. When the hotel manager left Burundi and returned to his home country of Belgium, Safari was surrendered to JGI. The traumas of his early life left their mark on Safari. He was easily excited and agitated. That made him potentially dangerous to people. I think Safari was about nine or ten years old by then. This is the time of life where male chimps move into adolescence, and they start to discover just how big and strong they really are. They're much stronger than a human by this point. And now I was locked inside a cage—just me and Safari.

The blows of a chimpanzee's fists and feet can be brutal. You can hear it when they pummel a steel enclosure door, slap the earth, or drum on the buttresses of a forest tree. They also have huge canine teeth and strong jaws, capable of removing a finger with a single bite. When they fight, the males often bite off the fingers—and sometimes the testicles—of their rivals. They bite each other on the face. They are very dangerous.

Trapped in the enclosure with Safari, I was surrounded by screams. Chimps were screaming out in the yard and inside the enclosure. The noise was incredible—so loud it was painful. Safari's hair was erect—a sure sign he was excited. He started jumping on me. He jumped on me many times. And he kept circling me, going around and around, swaggering and swaying and planning his next attack. He pushed me against the metal bars of the enclosure, which meant I couldn't escape him. I was scared, but I didn't scream out. I tried to stay calm, and I kept watching Safari, trying to decide what he was going to do next. Remember, this was in the early days of my work with chimps, and I didn't have much knowledge or training. But, I was lucky, because I had a small cleaning basket with me, and that gave me something to protect myself with. I tried to roll into a ball, and each time Safari came at me, I pushed the cleaning basket into his face, and that stopped him from hitting or biting me too badly.

After a while, I don't remember exactly how long, the screaming out in the yard grew quieter and then I heard Debby coming towards the enclosure. But still, I didn't call out to her. I just kept watching Safari. I was speaking calmly to him. He sat down. Debby came up to the enclosure and she started speaking to Safari through the bars as well. Eventually, the others distracted Safari, and I was able to move myself into a safe part of the enclosure and escape.

Debby took me back to the sanctuary office. I was bruised and feeling shaky, but I wasn't seriously hurt. Most importantly, I still had all my fingers and toes, and both my balls! I know Safari could have killed me, but he knew who I was, and I don't think he was really afraid of me. I don't think he considered me a rival. Maybe he was more excited than angry. Still, it was a very serious incident. I remember Debby gave me a beer after I was back in the office to help calm me down.

Safari is still alive. He is a little older and a little grayer now, and is a respected elder chimp, living in a sanctuary in Kenya. Sometimes, I wonder if he remembers the time he gave me that big pummeling. Maybe he does.

As far as I was concerned, though, that incident didn't deter me from working with the chimps. I had so much to learn. I was still very enthusiastic.

I really felt as if I had found my home with JGI at the halfway house. I was treated well and given better pay, and I was learning more and more about chimps. And there were exciting plans for the future. JGI Burundi wanted to move the halfway house chimps to a larger and more suitable sanctuary. They were interested in an area of forest near the Tanzanian border, and this was a potential location for a new, permanent facility. After they were moved to this new sanctuary, the chimps would be able to live out their lives without bars and concrete and collars and leashes, in a place that was safe from poachers and bullets and wire snares.

It was around this time that Doctor Jane Goodall came back to Burundi to visit the halfway house. I knew about Doctor Jane and her work, after hearing stories from Elie and Dean and Susanne and Debby. But I was not prepared for the excitement her visit would create.

People kept saying to me, "Hurry, we have to clean up and make everything look smart . . . Doctor Jane is coming."

I thought, wow, Doctor Jane must be like a government minister, she must be very, very important.

However, when we finally met in person, I found that Jane Goodall was so down-to-earth and friendly. She spoke to all the staff . . . and she even spoke in Swahili, so we could understand her better.

I remember sharing a cup of coffee with Doctor Jane one evening. She was so nice and easy to talk to, but there was something different about her. Do you know, when she had her coffee, Doctor Jane did not have any sugar in it. I couldn't believe it! In Burundi, we drank our coffee with a lot of sugar in it, but Doctor Jane, no, she had no sugar at all. I didn't know how she could drink coffee like that—even good Burundian coffee!

Oh, but she was very humble and kind and had a big influence on my life. I know that many thousands of people have been saved by her work—and even more chimpanzees.

For the first time since I'd left school, my future was looking positive. I was earning a steady income, and there was the promise of an exciting career at the new sanctuary. Perhaps too, at this sanctuary, I wouldn't have to walk thirty miles each way, every single working day. Perhaps I would live in a house close by. Yes, the future seemed really exciting.

But there were ominous clouds on Burundi's horizon.

It rains a lot in Burundi, so dark clouds are nothing new. But these clouds were different, because they were man-made. They were the clouds of explosions and fire and death. These clouds would create rivers of blood instead of water. These clouds would descend and destroy so much of the country of my birth. They would be the clouds of an all-encompassing darkness.

Chapter 3

APOCALYPSE LOOMS

It was April 1994, and near Kigali International Airport in Rwanda there were dark clouds on the horizon. An executive jet had crashed, striking the ground and erupting into a fireball. Everyone onboard died. The crash of any aircraft is tragic, especially when people are killed, but in this case, there was much more at stake, because the presidents of both Rwanda and Burundi were onboard the aircraft at the time. And this crash was no accident. Investigations would later reveal a surface-to-air missile shot the jet down. It had been a deliberate act and one that would have terrible consequences. Rwanda would face a bloody civil war and horrific genocide.

And then, there was my country.

The first truly democratically elected president of Burundi had been a Hutu man by the name of Melchior Ndadaye. Elected in June 1993, Ndadaye was killed in a Tutsi military coup a few months later. My country was heading towards civil war. Then, in the middle of this turmoil, the Burundian parliament elected another Hutu man as president in January 1994, when I was twenty-six years old. The new president's name was Cyprien Ntaryamira. It was Ntaryamira who had been traveling with the Rwandan president at the time their plane was shot down.

The deaths of Presidents Ndadaye and Ntaryamira would be the beginning of a catastrophe in my country. These two deaths led to the killing of hundreds of thousands of people. For the first time since 1972, Burundian Tutsis and Hutus were engaged in a major civil war. A tribal volcano, which had been smoldering for more than a decade, now erupted.

The conflict in Burundi was brutal. Clashes between government and rebel forces were unpredictable and horrifying. I try to forget the things I heard about—and the things I saw. But, even today, images of that war are enough to bring the nightmares back.

A group of soldiers weighed down by grenades, ammunition belts, and AK-47s, dressed in fatigue shirts and surf shorts, standing and smiling and smoking on the roadside beside a forest torn to pieces.

Hutu women, their faces cut and swollen from beatings, their eyes red from continuous tears, cradling naked infants too hungry and too exhausted to do anything more than breathe.

Visions of the smoldering remains of a village hut, where the silhouettes of charred human shapes are unmistakable in the ashes.

Long lines of refugees staggering through an angry, smoky heat, all earth-colored and astray, and holding onto each other as grimly as those who have lost everything but their lives.

All these things, we Burundians remember. I pray to God that one day . . . one blessed day . . . we will truly forget.

In Rwanda, it was the Tutsi people who suffered most of the deaths, but in Burundi it was Hutus like me and my family who were to experience the worst of this war. The government and military were mostly in the hands of the Tutsis. There were Hutu rebel forces fighting the Tutsi army, but for ordinary Hutu people like me and my family, there was no fighting—only the threat of death. All we wanted to do was live our lives each day. And we tried to keep living for tomorrow.

*

It now seems crazy, but in 1994, despite the chaos of the civil war, my life was filled with blessings. I'd been working at the JGI halfway house sanctuary for more than four years. I was earning a fair wage. I was learning more and more about chimpanzees. A new world was unfolding before my eyes. I really liked the chimps, and I found out they liked me too. They were becoming my friends.

This was a wonderful time for another reason. It wasn't just about my chimpanzee friends at the halfway house. There was someone else who had become special in my life. A girl called Nowera Nintegeka.

I can still remember the time I first met Nowera. I used to spend some time with one of my sisters, and she had a neighbor who she really wanted me to like.

My sister kept saying to me, "She's a nice girl, you know . . . and she is very kind."

Nowera was a nice girl, but there was a problem. She was only fifteen years old when we first met, and her family was very poor. My family wanted me to marry. I felt their pressure. As the eldest boy, I would bring a wife and children to the family village. Boys inherited land. Girls tended to move away with their husband's family after they were married. But Nowera was too young to get married. And my father wanted me to marry into a richer family. After that first meeting, it would be more than two years before I saw Nowera Nintegeka again.

During those two years, I met and proposed to another girl. I had given this girl traditional gifts—a dress, a watch, a necklace, and some lotion. I had to pay her family a dowry. All the money I could save from my work was going into this dowry. The relationship with this girl didn't last. Our wedding didn't take place. My parents didn't approve of this girl either, and I was heartbroken.

And then I saw Nowera again. My sister had never given up on me marrying Nowera. Two years had passed, and she looked very different. She had become a woman. I remember seeing Nowera put on a necklace, and I thought to myself how beautiful she looked. And she was a beautiful person on the inside as well. Even though she knew I had proposed to the other woman, Nowera didn't care. My parents also had a change of heart. Now they approved of Nowera. But there was a new problem. I had to save money all over again for gifts and for another dowry. That was hard, because I had spent all my money on things for the first girl.

I proposed to Nowera in 1990, but it would take me two more years to save up enough money to pay the dowry to her family. Because we were raised as Catholics, Nowera and I did not live together until the dowry was paid and we were formally married.

And it wasn't just the dowry I was saving for. I wanted my new wife and my future kids to live in a better house than I had grown up in. In Burundi, only the poorest people live under grass roofs. I had grown up in one of these grass houses and I really wanted my wife to live in a house with a proper roof, so I began to put aside a few schillings from each pay. My plan was to eventually buy corrugated metal sheets for the roof of our new family house.

Nowera and I were married in the local Catholic church in August 1992. This was a day of happiness in a country facing so much misery. We had a big party after the ceremony. There was lots of drinking, and we sang into the night! A cow was slaughtered, and we had a big feast. Nowera and I built our small house not far from my family home, and, I'm proud to say, our new hut came complete with a proper corrugated metal roof!

It wasn't just my personal life that was changing. Things at the halfway house were starting to change as well. As the unrest in Burundi was getting worse and worse, Dean and Susanne Anderson made plans to leave the country. Debby Cox, who had first come to the halfway house as a volunteer, had now been asked to come back and work in Burundi as the codirector of JGI, along with an American lady named Aly Wood. For me and Elie and the other staff at the sanctuary, the work continued.

Being married didn't mean an end to my long daily treks to and from work. However, there was one interruption to the routine. When I was first married, Dean and Susanne gave me a week of holidays to have a honeymoon with my wife. I know this would never have happened if I had still been working as a houseboy. Nowera and I put our time to good use! Our first child, a boy we named Charles, was born within one year of our marriage.

Meanwhile, the unrest kept moving closer to Bujumbura. Night after night, the soft, warm darkness would be shattered by the sounds of gunfire and explosions. And to me, it felt like the craziness and the danger was coming closer and closer with every shot.

The war was expanding, and, along with the other workers at the halfway house, I began to experience a direct impact of the fighting. On some mornings, I would be unable to safely negotiate my way to work. There would be some days when the presence of soldiers or the location of the fighting meant I couldn't make it safely out of the mountains.

I remember one morning when I was riding a bicycle down the mountain towards work. As I rode along, I started to be chased by a group of Tutsi boys. They were throwing rocks and yelling abuse. I was very afraid, because I didn't know what they would do if they caught me, or if they would go and report me to the police. So, I just kept riding. I had to keep going. I had to get to work.

Dean and Susanne were regularly being left alone at the sanctuary with all the chimps. When this happened, I would be worried. I worried about Dean and Susanne, and I worried about the chimps. There was no way two people could provide proper care for all twenty chimps, not by themselves. Important tasks couldn't be completed.

There was one particular occasion—I remember arriving at work for the first time in many days. On that day, I somehow managed to find my way out of the mountains without trouble. When I arrived at the house, Dean looked surprised to see me. He asked me how I'd managed to make it through.

I told him I didn't know.

Then, I went out to the backyard to see the chimps. They were in a terrible state. The big enclosure was especially bad. It hadn't been cleaned for a long time and the floor was wet and stinking and covered in mess. I could see worms and maggots, and they were creeping and crawling up onto the chimps. I was the only staff member who made it to work that day. I decided that, if I was the only one there, it was up to me to fix the problem. So, I went and collected my cleaning gear and I headed off to the enclosure to start my work.

Aside from my honeymoon and rostered days off, the thirty-mile trek to and from work was part of my daily routine. To get down the mountain and make sure I was at work by seven in the morning, I had to wake up at about three, while it was still very dark. I started walking about four and, because it was downhill, I could be at the halfway house before it was time to start. Walking home took longer . . . it was uphill. We'd leave at five in the afternoon, and I might not be home to Nowera until late at night. When I got home, I would eat some dinner, go to sleep for a few hours, and then get up and start the journey all over again.

I always believed that work was a blessing, especially when I was doing what I loved. However, things were becoming more and more difficult each day. As the war began moving into Bujumbura itself, just walking around the city became unsafe. For Hutu staff like Elie and me, going into the city to buy food or supplies meant risking arrest, beatings, or even death. But we had to do this, because we had to buy fresh food regularly for the chimps. In the mornings and at night, entering or leaving the city limits became particularly dangerous, especially if you were a younger Hutu man. The government soldiers were always looking for young Hutus, as we were the ones most likely to be recruited into the rebel forces. The more you moved about, the greater the risk you would be confronted by soldiers. And every single interaction with the soldiers came with risk. Some soldiers, you know, they were okay. They just did their duty. Others seemed less interested in being soldiers and more interested in killing Hutus. Some were in it just for the food and the pay. They weren't professional soldiers. The situation was unpredictable.

I was torn. I had a wife and a baby up in our village. The government soldiers and the rebels were raiding villages, looking for Hutus to recruit or kill. They could come to any village at any time, so I wanted to be at home to protect my family. But then, I also had responsibilities down at the halfway house. And I loved working with the chimps. I was caught between the two great loves of my life.

Still, even though I felt this pressure, I never once thought about quitting the sanctuary. Dean and Susanne—and later Debby and Aly—were told by the American embassy that they should consider leaving Burundi. But for Elie and me . . . well, this was our home. And we were never going to abandon the chimps. So, we tried to be smart about how we traveled to and from work. Before we started our journey, we would stop and listen to where the sounds of the fighting were coming from. And then we would try to plan a route that took us away from those areas.

But sometimes, no matter how well we planned, we would still be stopped by soldiers. "Identity card," they would demand. All Burundians were supposed to carry a national card, but it was especially important for Hutus, because this card had details of who you were and where you lived. If soldiers or a policeman caught you without the card, they would send you to jail. And sometimes, they would do worse.

There was one journey that none of us will ever forget. That was the day we took a car ride with Debby and Aly to the place where the road ended on our journey to the mountains. Events of that afternoon would not only change our lives, they would be the beginning of the end for the Jane Goodall Institute in Burundi.

When Debby first came to Burundi as a volunteer, Bujumbura city was mostly safe. Many of the homes and businesses in the city were owned by Tutsi people, which meant the soldiers, who were also Tutsi, left them alone. But then, right about the time Debby and Aly Wood arrived in Bujumbura to replace Dean and Susanne, the war suddenly became much more serious. I remember Debby's first week as the JGI Burundi codirector. It was a very memorable first week for her. And for all of us.

Debby had brought with her some equipment that had been donated by Taronga Zoo in Australia. There were new clothes and raincoats and other pieces of gear for each of us. We were grateful, but we didn't really like the idea of carrying all that extra stuff up into the mountains. So Elie and I, and another Hutu worker from the sanctuary named Ponpien, asked if we could have a lift in the sanctuary vehicle. Debby and Aly weren't happy about it but eventually said they would drive us . . . at least some of the way.

I remember the sanctuary vehicle from that time—it was an old Land Rover that was painted olive green. It looked like something that soldiers might drive. Maybe that wasn't such a good thing in a war zone!

Debby and Aly climbed into the front seats, and Aly was driving. Elie, Ponpien, and I climbed into the back of the car with our new gear. There were bench seats along the sides of the Land Rover, and we sat on those seats facing each other.

Debby told me later how uneasy she felt as soon as we drove out of the sanctuary gates. Aly said the same thing. There were no people on the streets and not even any animals. It was really unusual. They said it felt eerie.

Not long after we started driving, Debby and Aly stopped. They asked us to get out. They said they were going to turn around and go back to the sanctuary.

"Please, just a little further," we asked them. We said we were desperate to get back to our homes and our families.

They agreed to drive a little farther.

We didn't get far. We came across a roadblock. There were barriers and concertina wire stretched out across the road. I saw soldiers carrying AK-47s.

Aly stopped the car. One of the soldiers walked up to the window and pushed the muzzle of his gun inside. He pointed it right into Aly's temple. Aly gently reached up, put her fingers on the barrel of that rifle, and pushed it away from her head. When we told her about that later, she couldn't remember doing it.

The soldiers started shouting. There were lots of waving hands. Guns were being pointed. Looking out through the windows of that car was like watching a movie. It was chaos. It looked like anarchy. You could just feel everything slipping out of control.

Aly spoke French and a little Kirundi, and so she started speaking with the soldiers. She told them who we were, who we worked for, and that she and Debby were taking me, Elie, and Ponpien back to our villages to see our families.

Soldiers moved around the car. They opened the front doors and tried to drag Aly and Debby out, but the two women had their seat belts on and the soldiers couldn't get them out, so they gave up.

But they weren't going to give up on us.

The soldiers opened the back door of the Land Rover and they took hold of me and Elie and Ponpien, and they dragged us outside. I remember staggering out into the heat and the glare and confusion. I remember all the screaming. And I remember those guns. I couldn't stop looking at those guns.

They made the three of us lie down on the ground. We each had AK-47s pointed at our back. I can't remember another time when I felt so scared. I was lying facedown on the road, and I had no idea what those men were going to do. They could have done anything. We could do nothing. We were helpless as babies, just lying there on that dusty road while the boots of the soldiers shuffled and kicked and crackled on the ground around our faces.

Death hovered. You could just feel it.

We knew these men—these Tutsi soldiers—considered Hutu as some lesser class of people. We were barely human to them. And we'd seen how easily AK-47s could kill a human being, even from a distance. At point-blank range . . . we'd be mutilated.

One shot. Gone. There was no chance we'd survive.

And still, the soldiers kept screaming at us. At one point, I heard one of them say, "Hey, a bullet is too good for them! Let's just stab them."

Farther away, I could hear Aly's voice as she kept speaking to the soldiers. She kept saying that we worked for JGI and that we weren't rebels. She sounded annoyed with the soldiers, because I heard her say, "In America, soldiers are not allowed to do things like this!"

I was so scared. I kept wishing she would stop. I was sure these soldiers really wanted to kill us, and I thought she was only making them angrier.

It's hard to remember how long we were lying on the ground. Time doesn't move normally in these situations. But at some point, a more senior soldier . . . like a commander . . . arrived. Things began to calm down. At least the screaming stopped.

Elie, Ponpien, and I were ordered to stand up. The soldiers pushed us hard against the side of the car, and they pointed their AKs at our heads.

"Who are these men?" the commander asked.

"They fired at us from the car," the soldiers replied.

Of course, this wasn't true, but we had been riding in a Land Rover that was painted olive green and looked like a military vehicle.

The commander started to get really angry. The bullets had to be next.

Elie begged for permission to speak, but he wasn't allowed.

"How many Tutsis have you killed?" the commander asked us.

We didn't answer him. There was nothing we could possibly say that would help the situation, so we stayed silent.

The commander then spoke to Aly, I think it was in French, and he ordered her to turn the car around and drive back to the city.

We were going to be left there, alone with the soldiers. I thought that was a very bad sign. No Westerners now. No witnesses to report anything to the American embassy or any higher authorities.

The roadway was narrow where the soldiers had set up their road-block. Aly had to drive a little way down the road until she found a point that was wide enough to turn the Land Rover around. It seemed to take a long time for the car to come back.

No one fired. We were still standing.

When Aly and Debby returned, the soldiers suddenly lost interest in us. As Aly stopped the car, the commander said to her, "Here, you can take back your rubbish."

I can't tell you how fast we jumped back into the car! I remember seeing Aly's and Debby's faces as they turned around to look at us. I wondered if they had been crying.

They drove us to the beginning of the mountains and left us to walk the rest of the way home. When we got home, we told our families the story, and they were amazed. You see, most soldiers were only in the army because they were fed and given pay. And maybe some of them hated Hutus, I do not know. But they were not pro-fessional, disciplined soldiers like you have in countries like America or Australia. The Tutsi soldiers at that time could do whatever they wanted to Hutu men. There was nothing to stop them from killing us. Nothing at all. But, even after Aly and Debby had driven away and left us alone, those soldiers did nothing to hurt us. At the time I really didn't understand why this had happened, but today I believe we survived because of God. It was a miracle.

Meanwhile, Aly and Debby had safely driven back to Bujumbura. Debby later told me how the soldiers had yelled and waved at them as they drove past, like nothing had happened—like they were old friends. Debby also told me they had a few beers later that night.

The incident at the roadblock was a big deal for everyone at JGI halfway house—humans and chimps. Bujumbura was now busy with Tutsi soldiers, all hunting Hutu rebels. The JGI African staff were all like me, young Hutus. Tutsis considered caring for wild animals to be the lowest form of work, suitable only for a Hutu. This meant that any staff member could, at any time, be mistaken for a rebel fighter and dealt with accordingly. By this stage of the war, just being a Hutu in Burundi put your life in jeopardy.

To lessen the danger—and to reduce the length of our daily treks—Debby and Aly rented a room in a bunkhouse that was located much closer to the halfway house. This meant, on working days, we had to travel only a short distance to and from work. On weekends, or when we had a few days off, we were free to go back to our villages up in the mountains, if we were able.

Life in the bunkhouse room was simple. We had a saucepan, a couple of plates, and some cutlery. The average menu was simple and seldom changed. We would cook rice and beans for dinner. Breakfast was usually a donut and a cup of tea, and we were given lunch when we were at the sanctuary.

As the war continued to intensify, the daily commercial life of Bujumbura started to really suffer. It became harder and harder for us to find ingredients for even the simplest meals. During the day, merchants risked being caught in the crossfire between soldiers and rebels. At night, curfews were put in place, and no one was allowed on the streets. Commerce was crippled. Resources at the halfway house began to run short.

But you know, even when things ran short, the chimps were always given the best of whatever food we had available. We all made sure that those chimps never went hungry.

For the humans at the halfway house, things were less fortunate. We started to go without a lot of stuff. But still, even when there wasn't enough for the people to eat, we never considered stealing food from the chimps. Whatever food we were able to buy, we made sure the chimps got to eat first.

During the day we caregivers worked with the chimps, and at night, before the curfew began, we would go home to the bunk-house or up to our villages in the mountains. After we left, two armed watchmen would come to the halfway house and stand guard. Those two watchmen eventually paid a terrible price for their service to JGI and to conservation. One day, as they returned home to their villages, the two watchmen were ambushed by soldiers. Both were killed.

It was like the apocalypse had come to Bujumbura—the end of days. Fighting between government soldiers and the rebels moved closer and closer to the airport, which became a strategic prize for both parties. And, as the conflict came closer to the airport, it also came closer to us.

Burundi was now completely at war. The earlier plans JGI had made to move the halfway house chimps to a new forest sanctuary near the border with Tanzania were stopped. It was too dangerous. Fighting was everywhere. And there was no money. It started to look like JGI had no future in Burundi. A new plan was needed to save the precious chimps.

It was needed quickly.

Chapter 4

FREEDOM FLIGHTS AND HOMELESSNESS

By the mid-nineties, along with most of the Hutu population of Burundi, my family and I stood on the edge of the abyss. It felt like you might fall at any time. This would test us like nothing we'd faced before. What do you do when your whole country seems to be falling apart?

I remember speaking with Debby Cox at the end of 1994. She told me the halfway house was in real trouble. "We're right near the airport," she said. "This is good, because if we have to evacuate with the chimps, being close to the airport will be vital."

And, by this time the people at the US embassy were telling Debby and Aly it really was time for them to leave. The airport had become a strategic target, valuable to both sides, and the fighting was coming our way.

It wasn't just the halfway house at risk. The whole future of JGI Burundi was now in doubt. Debby said to me, "There's no way we can move the chimps somewhere else in Burundi now. This whole country is coming apart. Wherever we go will be really dangerous. And in any case, we just don't have money anymore."

The wider JGI organization decided that the halfway house chimps needed to be moved out of Burundi. There was a place in Kenya, a conservation reserve, near a town called Nanyuki. A chimp sanctuary called Sweetwaters was going to be established at this reserve, after agreement was reached between the organization that ran the reserve—called the Ol Pejeta Conservancy—the Kenyan Wildlife Service, and JGI.

At Sweetwaters Chimpanzee Sanctuary, we knew our chimps would have access to the sights and smells of wild Africa. The area covered two hundred acres, and there were real trees for them to climb and countless places to explore. There were birds and insects and other animals living in and around their home. All these things would stimulate the chimps' natural intelligence and curiosity. It would be a chance for them to feel the warmth of the sunshine and to breathe the fresh air. It sounded like a very good place.

But, we also knew there would be challenges for the chimps to solve in their new home.

"Most of these chimps have no experience climbing in naturally growing trees," Debby said to me. "Real trees aren't like climbing structures. A branch might look solid, but it can be rotten inside. It's going to be a nice experience for them, but they'll have to learn some important new lessons about tree climbing. Which branches will take their weight? Which branches might break?"

We were expecting some tumbles and falls. Still, at this new sanctuary, the chimps would be free from collars, concrete floors, and steel bars. At this new sanctuary they would be safe from human beings and their civil wars. And they would be able to live the rest of their lives in a proper chimp community.

Moving the ten adult chimps from the halfway house became our first priority. Debby and Aly would need time and resources to move these adult chimps to the new sanctuary. At that point, everyone thought moving the infants and juveniles would be the easiest part, because they were used to being handled, and this meant we could move them at short notice if we had to. But, when it came to the adults, things were more complex. For them, we had to make some serious plans.

While this was happening, Nowera and I had plans of our own. Family plans. We'd now had our second child, a brother for Charles, a little boy we named Innocent. You know, in the middle of the madness and evil of a civil war, this name seemed appropriate for a newborn baby. Still, we wondered, what kind of future our little baby would face. And what would be the future for his brother, and for the rest of us? What did the future hold for all the Hutu people in Burundi?

The big question for all of the team at the halfway house was, just how do you move a group of chimpanzees out of a city when that city is in the middle of a war zone? Debby and Aly knew we would have some trouble, but they didn't expect that government red tape would be one of them. We all thought the government of Burundi had more to worry about than moving a few chimps. We were wrong.

Debby and Aly were forced to work through the layers of bureaucracy. They had to complete a lot of paperwork. The chimps at the halfway house were considered Burundian, even though most of them had actually come from the Democratic Republic of the Congo. The government had to be convinced there was a need to move the chimps out of Burundi. So, JGI invited the Burundian foreign minister to come to the sanctuary and meet the chimps. During his visit, the minister would also meet Doctor Jane Goodall, who was in the country at the time. But it was the minister's encounter with one of our chimps that really helped things to move ahead.

I remember the foreign minister arriving at the halfway house. Mister Ngendahayo was a tall man, well-dressed in a smart gray suit, with a white shirt and blue tie. On his face he wore spectacles with round gold frames. Like most government ministers, he traveled with his advisers, but we were also in a war zone now, so he was guarded by two armed soldiers. Everyone walked around to the backyard, where the chimps were waiting for us.

As soon as they saw those soldiers, with their green berets, their camouflaged uniforms and those big AK-47s slung over their shoulders, the chimps were terrified. They started screaming and swaying and jumping up and down. Remember, most of these chimps were victims of poaching, so they knew what guns were.

Debby explained to the minister that "these chimps are orphans, and they have been taken from the wild by men with guns. Mostly they were babies when this happened. Each of them saw their mom killed before they were stripped away from her dead body. They are afraid of men with guns."

The minister immediately ordered his men to move out of sight, and once he did this the chimps calmed down. Debby and Doctor Jane then took Mister Ngendahayo over to see the chimps. They went up to the enclosure where the adult chimps lived. Poco, the alpha male who liked to meet new people, came over to the bars and reached out his hand towards the minister.

"Why is he doing that?" the minister asked.

Debby and Doctor Jane told him this was a gesture of chimpanzee greeting. Poco was thanking him for making the soldiers and the guns go away. And Poco kept reaching his hand out towards the minister. The minister reached out as well, and he took Poco by the hand, and then, as we all watched, the two of them shook hands with each other. The minister then took Poco's hand between both of his own, and he began to gently stroke Poco. Here was a meeting of man and chimp. Boss man meets boss chimp. Skin touching skin. It was a real connection. I remember seeing the look of joy on Mister Ngendahayo's face.

You know, after that, we got permission from the government to move the chimps to Kenya. I believe it was Poco who saved the day!

Even though we now had permission from the government, there were many hurdles to overcome. The US ambassador to Burundi was a man named Robert Krueger. He was a big help to JGI at the time. Thanks to the ambassador, the embassy offered us the use of a truck and assisted us to secure permits for the chimp move.

One of the main permits we had to obtain was a CITES permit. CITES stands for the Convention on International Trade in Endangered Species. Numerous governments around the world have signed this agreement to stop the trade of wild animals and plants. Chimps are endangered, so to move them from Burundi to Kenya required this permit, which meant a lot of paperwork. I remember Debby saying to me, "It's actually easier to move people than animals . . . even in the middle of a civil war."

Debby and Aly knew they couldn't do all the work on their own. Moving ten adult chimps would involve a team of people. And, Debby and Aly knew that we Burundians would be an important part of the moving team.

"Don't worry," we would say to Debby. "This is Africa. You just tell us what you need and of course we'll help. Don't worry . . . we'll make it happen for you."

While the paperwork was being finished, Debby, Aly, and the rest of us began to make plans for the actual move. No one had experience in an animal move of this size, and we had only limited resources. Space for the chimps was secured on a commercial plane, though not in the way you might think. Chimps may be closely related to people, but we knew we couldn't book them into passenger seats. Chimps are not very neat, and I hate to think what they'd do to the cabin of an airplane! So, we needed to follow strict procedures and use specialized equipment.

The most important of all was crates to transport the chimps. The crates had to be designed to fit into the cargo bay of a jet and be strong enough to hold an adult chimpanzee so it couldn't escape. We had nothing like this at the halfway house.

Specialized crates had to be made in Kenya. Then, once they were built, Debby needed to arrange permission and clearance from Burundian customs to import the empty crates into the country. After they had been imported, they had to be transported from the airport to the halfway house. Even though we were close to Bujumbura airport, the crates were much too heavy to be carried. They were made of steel, and each weighed 440 pounds—even without a chimp inside. They were also too big to be loaded into a car. The people at the US embassy arranged for the truck to go out to the airport, pick up the crates, and take them to the halfway house for us.

Things didn't go well. When Debby arrived at the place where the truck was parked, she found that the man who had the key had gone home. It was a Friday afternoon and Bujumbura was about to go into the evening lockdown. There would be a lockdown over the weekend as well. And the plane was due to leave on Sunday morning.

"If we miss this flight," Debby said. "I don't know when another plane will be available."

Eventually, the owner of the truck key was found, and they rushed out to the airport and the customs office.

"It was just before five in the afternoon," Debby said. "The customs office closed at five, and they were actually rolling down the shutters when we got there. We had to do a lot of begging, but we finally convinced them to open the doors, process the paperwork, and allow the empty crates to clear into Burundi."

Meanwhile, a second truck had been organized through the World Food Program. This truck would be used to take the chimp crates— with the chimps inside—back out to the airport on Sunday morning. When it arrived at the halfway house, this second truck became bogged in the driveway. We all had to dig and scrape and heave and push to get that truck moving again.

And then, after all that, Debby was told that the flight booked for the chimp evacuation had been delayed. "I really wonder if this isn't meant to be," Debby said. She looked dejected.

Debby Cox had been a trained vet nurse in Australia. She knew how to hand-inject animals. But, she'd never had to hand-inject an adult chimpanzee before, and there were ten adult chimps to be moved from the halfway house.

Chimps are so intelligent and perceptive. Even if they didn't know what a syringe was, as soon as they saw one of their friends injected and falling asleep, we knew the rest wouldn't let us go near them. The only way we could safely move the ten adult chimps into their crates was to first have them sedated. We didn't have a dart gun. Hand-injecting was all we could do. Debby told us we had to follow a strict process. We all had to work well together. We didn't have much time.

Once we had a new date for the flight to Kenya, we began to rehearse the procedure. Debby gave each of us a syringe filled with water and a papaya to practice on, so we all sat around and gave fake injections to pieces of fruit! But we knew it was important. We had to get it right on the day. We had to follow the exact process without making any mistakes. Debby had us all well drilled. We knew our jobs perfectly.

When the day of the move finally arrived, we were all ready before dawn. Because the halfway house chimps weren't trained to receive injections—like chimps are in many zoos—we had to come up with a method of injecting them without breaking their trust. I was learning how important trust was when working with chimps. If we broke the trust of these chimps, it would be harder to work with them again in the future.

So, this was our process: One chimp at a time, all the caregivers would make chimp grooming calls, smacking our lips together and inviting the chimp to come up to the bars for a grooming session. Meanwhile, the person with the syringe stood a little way back, with the needle hidden behind them. When the chimp came up to the bars, we would start to groom it. Once the chimp had settled, the person with the syringe would move up to the bars, reach through, give the chimp a quick jab with the needle, and then move away again. As they walked away, the injecting person would hide the needle out of sight again, and all the rest of us would make chimp alert calls. These are soft hooting calls that sound a little like *whooo... hooo-hah . . . whooo*. We did this because we wanted the chimps to think that we were as surprised as they were about what had happened. It was like we were saying to them, "Hey, we think something just bit you! No, sorry, we don't know what it was either. But we're here. We'll defend you. Don't worry." Because we did this, the chimps wouldn't associate the sting of the injection with any of the caregivers. This meant they would still trust us.

And the process was a major success. We sedated all ten chimps without losing their trust. After each chimp was sedated, one or two of us would go into the enclosure and the chimp would then be carried by hand out to an open crate waiting on the back of the truck. I still remember seeing Debby pick up one of the big males, holding him under both arms and carrying him out to the truck. More than a hundred pounds of chimp—and she was able to carry him on her own. Debby is amazing!

The chimps were gently laid into their crates onto soft beds of palm fronds and straw, so they would be as comfortable as possible when they woke back up and while on the flight. And we did it! Ten chimps were safely put into ten crates, ready for the short drive to the airport. There were no injuries to chimps or humans. The whole process only took us two hours.

Debby and Elie flew down to Kenya on that first flight. The plane was a full-sized jet. Debby could go down into the cargo area because the cabin was pressurized, and that way she could regularly check on the chimps.

As the plane was coming in towards Nairobi, the pilots asked Debby to come up into the cockpit. "I remember how clear the day was," she told me. "The flight path was between Africa's two great mountains. Mount Kilimanjaro was on one horizon, and Mount Kenya was on the other. It really was so beautiful."

When they arrived in Nairobi, the ten chimps from the halfway house were loaded onto a truck and driven 130 miles straight to

Sweetwaters Sanctuary. Because there were no chimps living wild in the Kenyan bush and there were no chimps living at the sanctuary at this time, the Burundi chimps were allowed to be in quarantine at the sanctuary itself. Once time had passed and they were used to their new surroundings, Poco and Socrates and Safari and the other adults were allowed outside. Protected by tall electric fences, our chimps were free to explore the trees and the grasslands. It was the first time they had been outside in the sunshine and the fresh air for many, many years.

Meanwhile, I was still back at the halfway house. Even though the war was closer and closer to us, we were all feeling happy. The first part of the job had been completed. The adult chimps were now safe. The most difficult part of the move had been a big success.

We thought that moving the adult chimps was going to be the hardest part of leaving Burundi. But we would soon discover that moving the babies and juveniles was going to be challenging as well.

It was the end of 1995, and the war was everywhere. Bujumbura was in chaos. The airport had been attacked so many times that the facility couldn't be used by big aircraft anymore. How were we going to fly ten young chimps to Kenya? Perhaps we would have to sail down Lake Tanganyika into Tanzania and then fly from there to Kenya?

However, there was a man named Russell, who owned the company that ran Ol Pejeta Conservancy, where the Sweetwaters Sanctuary is located. Russell was friends with Doctor Jane, and so he donated his private plane to evacuate the younger chimps. This was wonderful, but it created a new set of problems. Even though the juvenile chimps were used to being handled by humans, now they were going to be transported in a small plane. If they escaped, they could easily get to the pilot of the plane. Even as juveniles, chimps are very strong. They could easily hurt the pilot and the plane could crash.

"I wasn't worried about the second move until we started to do it," Debby said to me.

Because there was not as much space on this smaller plane, the younger chimps couldn't travel in large crates, so they were placed into plastic pet packs for their flight . . . you know, those plastic carriers you might use to take your dog or cat to the vet. The chimps weren't sedated for the flight either. I think Debby was worried about what might happen!

Later, Debby told me that second flight to Kenya was a long one. She had some syringes of ketamine ready to sedate any chimp that escaped from its pet pack, but they all behaved themselves very well on the plane. "They were like ten little angels," Debby said.

But, after they arrived in Nairobi, Debby's ten little angels started to become naughty. Because there were now chimps living at Sweetwaters and because many months had passed since the first move, the ten younger chimps had to stay in quarantine for a month in Nairobi before they could join the adults out at the sanctuary. The quarantine took place at the Nairobi Animal Orphanage. When they arrived in the orphanage holding area, the ten juvenile chimps began to bust out of their pet packs. And all ten of them went running straight towards Debby. They all jumped on her. I think she must have looked like a walking tree, poor Debby, with ten little chimps all hanging off her.

The day after the juvenile chimps were flown to Kenya, it was my turn to leave. I wondered how long I would be away from Burundi— and away from Nowera, Charles, and Innocent. I was worried about leaving them at such a dangerous time, but it was my job to work with the chimps and help to settle them in Kenya, so I had to go. And I wanted to go as well. I loved the chimps, they were also my family.

I flew down to Nairobi on a small plane, like Debby did. But, unlike Debby, I had never been on a plane before. I was so nervous. And as we sped down the runway faster and faster and then lifted up into the air, I was very scared.

Of course, I had seen planes fly overhead many times. Working near Bujumbura airport, there were many planes flying around. But, when you see a plane high in the sky, it looks smooth as it flies along.

I found out it isn't like that when you're inside. As we were flying along, the plane started shaking. I didn't understand why this was happening. You can't see turbulence, and I didn't even know what that was! I was so afraid. I was sure we were going to fall out of the sky. Landing on the ground in Nairobi was a big relief.

Elie had traveled with Debby and the ten adult chimps to Kenya, but he had come back to Burundi after about a month. My stay in Kenya was to be much longer. After they had finished in quarantine, it was my job to integrate the ten juvenile chimps into the adult group now living at Sweetwaters Sanctuary. I didn't know it at the time but leaving Bujumbura on that small plane was to be the beginning of many great adventures. My life would take me to places far beyond the mountains of northern Burundi.

And the first big adventure would begin when I arrived at my hotel in Nairobi.

Bujumbura is the largest city in Burundi, but it's not as big as world cities like New York or London or Sydney. So, even though I knew a little about city life from Bujumbura, I really wasn't prepared for what I would experience in my first truly big city, Nairobi in Kenya.

Debby had arrived in Nairobi the day before me, traveling with the ten juvenile chimps from the halfway house. To save money, she had booked a single hotel room where the two of us would spend the night after I arrived. Now, I was a married man, so I want you to know that the room was double-size and we had separate single beds. No funny business!

After I arrived in the city, I met Debby in the hotel restaurant and we had dinner together. Once we'd finished dinner, Debby went up to the room, and I followed a little bit later. Debby told me the floor number and the room number, so that I knew where I was going.

I didn't know about elevators . . .

I went up to the elevator, and I pressed the button and the doors opened. I went inside and I pushed the button that had the floor number Debby had given me. The doors closed, and the elevator started to move. Then, when the doors opened again, the first thing I saw was a man in uniform, holding a gun. I was shocked. I let the doors of the elevator close, and I stayed inside and rode back down to the ground floor again. I was sure I had gone to the wrong place. Why was there an armed man standing there? He looked like a soldier, and, after living in a war zone for so long, I didn't trust soldiers. I was also scared to get out of the elevator. What if I walked out and the doors closed while I was still outside? I wasn't sure if I could make the elevator come back to me again. And if that happened, I would be trapped with a man holding a gun. When the elevator went back to the ground floor and the doors opened, I pressed the floor number that Debby had told me, and the doors closed again. When the doors opened . . . there was that soldier! I was confused. I was afraid. Where was Debby? What was I going to do about that soldier? I let the doors close again and went back to the ground floor.

I did this three more times.

The fourth time, I pressed the number and the doors opened and I saw the soldier again, but this time he yelled out, "Hey you, why do you keep coming and going? Why won't you get out?"

I told him I was scared to get out. I didn't want the elevator to go and leave me because maybe I couldn't make it come back. But, this time I did get out of the elevator, and I walked to the door with the number that Debby had told me. I was still worried I was in the wrong place and was lost. But I knocked, and I was very glad when it was Debby who opened the door!

And do you know, I never told Debby anything about my adventures in that elevator. I just got straight into bed and went to sleep. After she reads this, she will finally know what happened. And I'm sure it will make her laugh!

There wasn't much to laugh about at the Nairobi Animal Orphanage. The juvenile chimps from Burundi, including my favorite boy, Max, were in need of a friend. The orphanage is only a few miles from the center of Nairobi city and receives thousands of visitors each year. Chimpanzees are not native to Kenya, and so many of the people had never seen one before. Because the visitors were so curious, and because the chimps seemed so human, some unhelpful things started to happen. Some visitors started teasing the chimps. They threw sweets and cigarettes at them. The chimps started to get nervous and scared. So, I began sitting with them in their enclosure. I wanted the chimps to know they had a friend with them. And I called out to the visitors and asked them to stop throwing things into the enclosure.

There were other challenges for the Burundi orphans. Most had been born into the warm tropical rain forests of the Congo. Bujumbura too has a tropical climate, and even at night the temperature rarely falls below twenty degrees Celsius (sixty-six degrees Fahrenheit). However, even though it's in the tropics, Nairobi sits at almost six thousand feet above sea level, and the nights can be cold. The chimps didn't like it much.

When I arrived each morning, I would find them looking miserable. I felt so sorry for them. But, when they saw me arrive, they would call out their welcomes. To do this, chimps pant-hoot. These are long hooting calls broken by loud intakes of breath. They sound like, *whoooo-haah . . . whoooo-haah . . . whoooo-haah . . . whooo.* When they made these calls, I knew they were happy, because they were saying to each other, "Hey, Stany is here. Wake up everyone . . . here comes our man . . . here comes our dad!" This made me feel good!

When they had completed their time in quarantine, I went with the ten juveniles out to Sweetwaters Sanctuary. The plan was to unite them with the adults from the halfway house to form one integrated chimp community.

When we arrived, the smaller chimps were put into an enclosure, away from the adults. They were nervous. They kept pacing around their enclosure, pant-hooting to announce that they had arrived. When they did this, the older chimps started pant-hooting back. As soon as they heard this, the juveniles began to calm down. The adult chimps were welcoming them. Those pant-hoot calls were greetings. When I heard the two groups calling to each other like that, I knew it was going to be okay.

The twenty chimps from the halfway house became two-time refugees—first as orphans from the forests and then from Burundi where a human war meant they had to leave. Finally, at Sweetwaters they found a permanent home. And today, more than twenty years later, many of those chimps still live at the sanctuary, including Poco and Max, my first chimpanzee friends.

I was, however, now a man alone. And I began to wonder what my future would be.

Not long after I arrived at Sweetwaters, Debby Cox left Kenya. She had been offered a position with JGI in Uganda. I was told she would only be away for about a month, but that time began to increase. Debby left some money for me to buy food and to look after myself, but apart from the chimps, I was now living alone.

It was at this point when my life began to take an unhelpful turn. It wasn't just being on my own, because I did have the chimps for company, but I was very worried about things back home. I didn't know much about the situation in Burundi, but what little things I heard upset me. The fighting was getting worse and worse. I had a wife and two little boys living there. The rest of my family was there too. I didn't know if any of them were safe or if they were even alive. I wanted to be with them, but I couldn't go back. Meanwhile, Debby had left me and gone to Uganda.

I stayed with the chimps in Sweetwaters week after week. Each day, I would walk four miles into the local markets to buy extra food for them. I used the money Debby gave me for my food and also used it on the chimps. I was happy to go without some food to make sure my friends were happy and healthy.

Then, at night, I would start to feel very lonely. I thought that maybe alcohol could be my friend. Babies in Burundi, when they cry a lot, they are given banana liquor to help them sleep. So, I knew that alcohol would calm a person down. And I believed that alcohol would make me feel better. You know, to begin with, it really did. Drinking took my mind off my problems and helped me not to worry so much about my family.

After four months, we felt that the chimps had settled into life at Sweetwaters Sanctuary. My job there came to an end. So, I began looking for work in Kenya. I couldn't find any. I was a Burundian Hutu, and there was nothing for me. I was unemployed and worried and sad. I decided then I was going to go back to Burundi, just like Elie had.

I spoke to Debby, and she told me I shouldn't go back. "You are a young Hutu man," she said to me. "You're the right age to be considered a rebel fighter by the Tutsi soldiers. If you go back, you'll be arrested as soon as you try and cross the border. They will put you in jail. They might even kill you. Then you will never see your family again. Perhaps, one day, there will be another way to unite you with your family, but right now, you can't go back. Not the way things are."

Though I didn't know it, my time in Kenya was about to come to an end. Debby had other plans for me. I would move to another country and into another job. This new job would change my life completely. This new job would help to save my life. There would still be dark times, but I would eventually rediscover forgiveness and faith. And I would spend more and more time learning about my friends, the chimpanzees.

This was going to be the time of the Chimp Whisperer.

Chapter 5

COURTING DEATH– CHOOSING LIFE

It was 1996. I was stateless, homeless, and far away from my family. Worse, my family was trapped inside the most barbaric civil war. My world felt hopeless. Because I wasn't working with the chimps at Sweetwaters anymore, during the day I would try to go out and look for some work, and each night I would go out drinking. I was smoking too much. I was drinking even more. I was a man who was slow dancing with death.

Debby Cox had been invited by the Jane Goodall Institute to go to the Entebbe Zoo in Uganda. JGI wanted her to investigate options for some chimp projects in that country. But, by this time, Debby had made plans to go back to Australia and to her job with the chimps at Taronga Zoo. She wasn't sure about the political situation in Uganda. She had just left Burundi, where genocide and civil war had created so much misery and hardship, and she didn't want to go through all those things again.

Uganda too had seen years of fighting and massacres led by dictators like Idi Amin and Milton Obote. Still, Debby decided to go and see for herself. She visited Uganda, and she spent time speaking with the locals, and it was these people who gave Debby a feeling of hope. They told her they'd had enough of war and suffering and that they wanted to live normal lives again. They just wanted to live in peace. When Debby heard this, she decided she would delay her return to Australia and spend some time working for JGI in Uganda.

JGI had a vision to create a new chimpanzee sanctuary somewhere in Uganda. But, there was no money, no land, no resources, nothing. There was, however, a vision, and that was really all someone like Debby Cox needed. She would go on to become the executive director of JGI Uganda, a role she would fill for many years.

One of Debby's first tasks was to consider the kind of people she might need to help turn the JGI vision into a reality. Later, she told me, "You know, as soon as I began thinking about making a fresh start with chimp projects in Uganda, I thought of you. I trusted you with chimps. I trusted you with my life. I knew you wouldn't let me down."

Five months after she left Kenya, Debby arranged papers for me to come to Uganda as a refugee. Debby warned me there was no permanent job at that stage, and I didn't really want to go to Uganda. I had been told it was not a good place for Hutu people to live. There were many refugees living in the country by this time, and many of them were Tutsis. And, some of the local people didn't like refugees at all. I felt torn, but Debby said that I should go. I trusted Debby. I loved Debby. She was like my sister. And then she told me that she would arrange for my brother, Minani, to come to Uganda with me, so I decided that I would go.

Those early days in Uganda were tough. Leaving Sweetwaters, and especially leaving Max and Poco and my other chimp friends was difficult. Aside from Debby and my brother, I didn't know anybody in Uganda. I had no real friends. I was still drinking. It wasn't long before Minani and I began drinking way too much.

Because we were refugees in Uganda, every three months Minani and I had to travel from Entebbe to the capital city, Kampala, to visit the United Nations offices there. We would stand in long queues waiting to reapply as refugees. It was hard. It was also a little bit dangerous. There were many refugees living in Uganda by this time. Many were Hutu, like Minani and me, but some were Tutsi. Sometimes we would see them looking at us. I could see the hate in their faces. I was worried that these men would be prepared to kill Hutus if they had the chance. Even though we were now safe from the fighting, there were many people who still wanted revenge. That made me very worried.

At the UN, we could see firsthand the terrible toll that civil war and genocide were having across Africa. We stood in those lines with refugees from Rwanda and Sudan and Congo, as well as with other Burundians. We had to wait for a long time.

When it came to be our turn, we had to speak with some people from the UN who would interview us. Many of the other refugees wanted to get out of Africa. They wanted to go to the United States or Canada or the UK or Australia. Minani and I told the UN workers we wanted to stay in Uganda. I didn't know if that was a good thing or not. The future felt very uncertain.

In times like these, where life is blurred by so much uncertainty, I believe people need a firm foundation in their lives. We all need something or someone to offer support and comfort. For me, at that time, my foundation was Debby Cox. Debby was always there for me. She was so kind. And, you know, she gave Minani and me military training. She had us very well drilled. For many Africans, time is not so important. But it was to Debby! She would make sure we were on time for everything. She was strict with us and made sure we went to Kampala to renew our refugee papers when we were supposed to. Debby would say the word, and off we would go!

Eventually, Minani and I were given Ugandan citizen cards. Once we had citizenship, we didn't have to travel to see the UN anymore. Oh, that was a very good day! Minani and I began to settle into our new life in Uganda. Debby hired Minani as her housekeeper, and I began a new job working with Debby and the chimpanzees of the Entebbe Zoo.

The early days at Entebbe Zoo were tough. The chimps themselves were in good condition, but the enclosures at the zoo were really old and not really suitable for chimps to live in. It took many weeks to clean them all out. We worked away with brushes scrubbing through layers of dirt and shit and grime just to get down to the original paintwork.

Debby also had adventures with some other residents of the zoo. She had a colleague with her from Taronga at the time. His name was Anthony. Debby and Anthony would spend entire days doing what she called "rat bashing." This involved closing themselves into a store-room that was infested with rats. They both had big sticks with them, and when they had the rats locked inside, they started swinging those sticks. They might kill twenty rats at a time.

To start with, many of the staff at Entebbe Zoo didn't like having Debby and me working there. Debby said to me once, "You know, we are on their turf. We have experience, and many of them don't. They might think we are here to take over."

I'm sure those staff were worried that they might lose their jobs. If I had been in their shoes, I think I would have felt exactly the same way.

But I knew Debby wasn't like that. She always wanted to work cooperatively with African people. She didn't try to impose her ways on any of us.

To improve the technical skills of the Entebbe Zoo staff, and to help us all operate at the same level of expertise as zookeepers in other countries, Debby thought we should complete some sort of formal education. In Australia, zookeepers completed a course called Certificate in Animal Management. Debby took this course and rewrote it to make it suitable for African zookeepers. Then she set about training all the staff at Entebbe Zoo.

For me, completing the course was going to be a big challenge. I could speak some English but not so much. Debby and I had always communicated using a combination of English and Swahili words. I couldn't read or write in English at all, and that was a problem, because all the training materials from the Certificate in Animal Management course were written in English. I was worried that if I couldn't complete the course, I would be even more isolated from the other staff than I already was. I went to Debby and asked what we should do. She came up with a solution.

I could understand Debby when she spoke with me, so she read out all the coursework. She read it out only once. Some of the information from the course I wrote down in Kirundi or Swahili, and I kept asking Debby questions. Even after we had finished work, I would keep asking her questions. Some of the other keepers spoke very good English, but they did not do so well in the course because they weren't especially interested. But for me, this was my love, and so I put all my heart into it.

After we had finished the course material, Debby told us we had to complete an examination. The keepers who could read and write in English were able to do the exam on paper, but I couldn't do this. So, Debby read out the questions to me, and I answered them out loud. And then she wrote down my answers in English. I passed the Certificate in Animal Management course on my first try. I was very pleased to achieve this, because I had done only three years at school and I knew that some people—even in Australia—didn't pass this certificate course.

However, not all the tests I faced in Uganda were about study and exams. There were other kinds of tests going on in those early days at Entebbe Zoo.

Many Ugandans didn't like being told or shown what to do by a Burundian. And Hutus were often considered to be poor people who only did jobs like houseboys or gardeners. This meant I had trouble working with some of the other zoo staff, even though I had passed the animal management course and even though I had some good experience working with chimps.

At first, the tests were only small. The enclosures at Entebbe Zoo were in poor repair, which meant the strong and clever chimps would sometimes find weak spots in the wire. If they worked away at the weak spots, sometimes they could make a hole big enough to escape through. And when one of the chimps escaped, all the other zoo-keepers would run away in the opposite direction. They would leave me to deal with the situation by myself.

I still remember one particular escape. The other keepers ran away, but I started running in the direction of the enclosure. When I got close, I could see that one of the adult male chimps was trying to climb out through the wire roof of the enclosure. So, I climbed up the side of the enclosure and then, once I was at the top, I crawled along the wire until I came to the point where the chimp was trying to escape. I reached out to the chimp and I began pushing him back down inside the wire again. It's funny, because I can still remember saying to him, "No . . . you . . . will . . . not . . . escape . . . today." And, you know, each time I pushed down on him, another word was pushed out of my mouth. Eventually, some other keepers came to help me, and once the chimp was back inside, we got some more mesh and we replaced and reinforced the top of the enclosure.

Then, the tests became more serious. They became life-and-death serious.

I was standing with some of the keepers—including the head keeper—when they started saying to me, "We've been told you are a great chimp keeper."

"I'm okay."

"Well," the head keeper said, "any great chimp keeper should be able to go in with these chimps and feed them." He then gave me a bucket of food, and he ordered me to go into the enclosure.

I didn't think this was the right thing to do. I knew how dangerous chimps could be. I was also sure there were rules against this kind of thing. But he was my boss, and so I did what he told me.

The chimp group at Entebbe Zoo was large, and while most of them were juveniles, they were still strong and potentially dangerous. There was also one big adult male, called Robbie. Robbie was a very big chimp. He was capable of killing a human if he wanted to. And I hadn't forgotten the pummeling I received from Safari back at the halfway house. Still, I went into the enclosure like the head keeper told me to. I didn't have anything for protection, and I was carrying a bucket full of food.

Once I got inside the enclosure, my heart started pounding so hard I could hear it in my ears. The chimps started making loud alarm calls, *whooo . . . whooo . . . whooo*. They were banging on the floor, and their hair was standing up. I was watching the chimps really closely . . . watching . . . watching. And I kept my eyes on Robbie. He was the leader. If the group was alarmed, I knew he would defend them. But if he stayed calm, then they would stay calm as well. I knew I had to make friends with Robbie.

I started spreading food around the enclosure. This took the attention of many of the chimps, and they stopped making alarm calls. Then, I looked into the bucket and found the biggest and best banana. I offered it to Robbie. As I did this, I made chimp submission calls, which sound like a soft, *haa haa haa*. Robbie came at me, ready to bite me, but I offered him the back of my hand as this is another gesture of submission for chimps. I wanted to show Robbie that, in this place, he was the boss of me. I wanted him to know that I respected him. And he didn't bite me.

After a while, Robbie walked up to me until he was very close. He looked into my face and then he turned around and sat down and offered me his back. He had accepted my gesture of submission and he now wanted me to groom him. Grooming is a way chimps build relationships with each other. This was good timing, because there was another male in the enclosure called Sami. He was still unhappy with me being in there. Sami was calling and trying to encourage the other chimps to attack me. But I now had the alpha male, Robbie, as my friend. We were grooming each other. So, Sami didn't dare attack. This was a good lesson—making friends with the boss chimp is always essential!

Eventually, the food ran out, so I put the bucket down on the floor of the enclosure, turned it upside down, and I sat on it to show the food was all gone. Sami's attempt at inciting an attack was over. The other chimps started approaching me and Robbie, and they started asking if they could also be groomed. They made squeaks with their lips, like raspberry sounds, which chimps sometimes do when they want the attention of another. I had been accepted into the group, so I made sure that I groomed every one of those chimps before I left the enclosure.

Twenty minutes later, the other keepers let me out. Not only was I unharmed, but I had made friends with the alpha male and the rest of the chimps. I think then some of the keepers could see that I was able to communicate with the chimps. It wasn't just talk. However, acceptance from the keepers would be harder to win. They started rumors about me. Some of the keepers spread the idea that I had practiced witchcraft on the chimps and that was the only reason they hadn't attacked me!

And, you know, Debby found out what had happened a few days later. I won't repeat the exact words she used, but she was really furious. She brought all the staff in and gave everyone a big lecture. There was yelling. She told us she couldn't believe we would do something so stupid.

There was another serious incident at Entebbe Zoo. This time, the boss of the zoo was involved—as well as the boss of the chimps. Robbie had escaped from his enclosure again. He was displaying, running around with his hair up, making himself look big and tough. And while he was out, he charged Wilhelm.

Wilhelm Moeller was a very short man, with a sharp, seri-ous-looking face and handsome silver hair. He had been the director of Entebbe Zoo since 1995. With the help of people like Debby Cox, Wilhelm would be celebrated as the man who transformed Entebbe from an old-style concrete and steel zoo to a world-class conservation facility that would be renamed the Uganda Wildlife Conservation Education Center.

Robbie didn't seem to care about any of that. He didn't care that Wilhelm was the boss. When he came to Wilhelm, Robbie hit him hard, and this knocked Wilhelm over. Robbie jumped on him. When I arrived at the scene, Wilhelm was pinned to the ground. It was a serious situation, and though he remained calm, I'm sure Wilhelm must have been very afraid. I approached Robbie slowly. I was speak-ing to him calmly and softly, and as soon as he saw me, he climbed off Wilhelm and walked over to me. When he got to me, Robbie turned around, sat down, and invited me to groom him. We were friends again. After this happened, Wilhelm was able to get back onto his feet and escape.

After a short time, Debby arrived with a syringe filled with ket-amine. Robbie moved away again, so Debby and I started making the chimp alert calls. We did this to distract Robbie and encourage him to come back to where I was. Even though he was an alpha male, out-side the enclosure, Robbie was nervous. He had a wide grin on his face, with his lips pulled back and his teeth showing. This was a fear grin, and chimps make this face when they are uncertain or afraid. I knew Robbie would feel safe if he was with me, and when he heard Debby and me calling, he came straight back. Debby hid the syringe behind her back, and when Robbie turned around to be groomed, she gave him an injection of sedative. After he went to sleep, we were able to carry Robbie back into his enclosure. Wilhelm wasn't seri-ously hurt.

*

Away from work, my life was not so good. Aside from Debby, Minani, and the chimps at Entebbe Zoo, I had no friends. I felt lonely and depressed in Uganda. Life felt like an open wound. The people in Entebbe didn't know what I had been through, and the attitude of some of them to me only made my wounds deeper. My wife was far away. Or maybe she and my little boys were dead. Whenever and wherever I could, I would listen to the BBC World Service on the radio. Sometimes there would be reports about the situation in Burundi. The news always seemed to be bad. The reports said that many people were dying. I was far away and lonely, and I was afraid for my family. This was the reason I started to drink heavily.

I had been drinking and smoking since I was a young man, but after I left Burundi, it became different. I started drinking more heavily when I got to Kenya. You know, I really thought it was helping me to feel better. And then, after I went to Uganda, I met a man at Entebbe Zoo, and he invited me to join his drinking club. We would meet at a small bar or sometimes in peoples' houses, and we would try the local drinks. The people there spoke to me in Swahili—many Ugandans didn't speak Swahili—and this immediately made me to feel more at home. So I kept going along to join in and drinking every night.

In many African cultures, drinking is much more than a social pastime. It isn't just a chance to catch up with your friends. When you drink, you do it to get drunk. The drinks in Uganda were spirits made from millet or fermented banana and sugar, and they were called Uganda Waragi. These drinks are really strong. No, more than strong, they are like petrol. And drinking this stuff was harmful. It would only take a few glasses and you were down. I was smoking a hundred cigarettes during the day and then getting completely drunk every night.

They say realization is the first step on a journey to healing, but even when I realized I had a drinking problem, I didn't stop. I was wasting all the money I made at the zoo on drink and cigarettes. But, you see, I still thought I would go back to Burundi one day. And thinking this way made me miss my family more. And the more I missed them, the more I drank. The more I drank, the more money I wasted. It went on and on.

Debby was aware of my problem. She knew I was falling into a deep hole, and she tried her best to stop me. But, even though I loved her and was grateful for her support, I still went out and got drunk almost every night.

Then, later that year, I met a woman who ran a health clinic in Entebbe. I was at the clinic to buy some medicine for a friend. The woman started talking with me about the church she went to. She said to me, "You know, there is no medicine that can work without God."

I had grown up a Catholic in Burundi. I had even been to the Catholic church in Uganda a couple of times. I knew about God, but the church this lady spoke about sounded different. I knew I needed help with my drinking. My life was becoming very bad. I was interested in going along to this church and asked my supervisor if I could have Sundays off. But, Sundays were really busy at the zoo, as they are in zoos around the world. More visitors came to the zoo on Sundays than any other day. Chimps were still escaping from their enclosures on those days, and my supervisor wanted me to be there in case a chimp escaped, so he wouldn't let me have that time off. I suggested I would only take one day off each week, instead of the usual two, if that day could be Sunday.

The next Sunday, I went to the Entebbe Miracle Church with the woman from the clinic. It was so very different from what I had experienced in the Catholic church. It was the most amazing church service, filled with sound, color, and movement. It was full of life and love and joy. They asked people to come forward if they were new. So, I went up the front, and I told my story and everyone clapped and cheered. That had never happened before. There were people dancing and hooting, which sounded a lot like chimpanzees' pant-hoots, and I felt right at home when I heard the people doing that.

Then, it was time for the sermon.

I remember they were preaching about the bad things in life. You know, by this they meant sins. And when they spoke about sins, they started speaking about things like drinking and smoking. I listened to them speak. I really listened. After hearing that sermon, I told myself the way I was living was wrong. I knew my brother Minani was going down a bad path, and I didn't want to follow him. I had responsibilities—to Debby and Wilhelm, to the zoo and to the chimps—so I made a prayer. I was determined to stop. I prayed to God and asked Him for help. And after that prayer, I stopped drinking and smoking.

From the day of that church service, I completely stopped.

I was excited to tell Debby and the others about my newfound life. Some conservation people have a negative view about God. I mean, some were okay, but others spoke bitterly about going to church. I didn't really know why anyone would be negative about my decision. My work got better and better each day. I was coming to the zoo with a clear head, more energy, and no hangover. So, even if people didn't like my choices, what could they say? There's no doubt in my mind that because of God and His church, my life began to get better.

One of the other benefits—although I consider it a blessing—of going to this new church was that services were delivered in both English and Luganda—the local language spoken in Entebbe and Kampala. The more I went to church, the better my English became, and I also learned the local language. And there was more, I started to exercise and look after my physical health as well as my mental and spiritual health. Life turned a corner.

But, even in the house of God, there were still cultural and tribal divisions. There was a Rwandan woman in the church; she was a Tutsi and she knew I was a Hutu. She started telling people that I had been a rebel soldier in Burundi and that I had killed many innocent Tutsis. It wasn't true, of course, but it hurt my feelings. I was a Hutu man who had stared into the gun barrels of hatred, and now I was being accused of the same atrocities. Still, I didn't retaliate. I didn't seek revenge.

I am fascinated by the story of Joseph from the Book of Genesis. Joseph was the beloved son of Jacob. He was betrayed by his brothers, but when he had the chance to seek revenge, instead he forgave and reunited his family. In the same way, I believe in the power of forgiveness. I don't want to live a life directed by revenge or hate. I don't believe that is God's way, so I won't let it be mine.

One of Debby's major ideas was to work with Wilhelm Moeller on the creation of a "chimp island" exhibit at Entebbe Zoo. Based on the plans used to construct the chimpanzee park at Taronga Zoo in Sydney, it would be an open-air enclosure, surrounded by a water-filled moat, with an enclosed raceway connecting the new exhibit to the existing chimp facilities. The problem, as is often the case, was a lack of money.

Eventually, thanks to a donation from the World Bank, the enclosure was finally constructed. Surrounded by a deep water-filled moat, more than twenty of the zoo's chimps had the chance to leave their old wire enclosures and enjoy the feeling of soft grass under their feet. They could climb real trees and swing on ropes and tires hanging from those trees. It was so wonderful to see them out in the fresh air.

And I remember, not long after that new enclosure had been opened, Debby and I went out to have dinner to celebrate. "Now Stany," Debby said. "What more can we do?"

"I'm not sure," I replied.

And then Debby said something I will never forget, "We need to build a proper home for the chimps. Maybe a sanctuary . . . out on an island!"

Soon, there would be a chimp sanctuary unlike any other in Africa at the time. It would be visited and celebrated by animal lovers and researchers from around the world. It would be a model for others to follow.

For me, though, there were many unanswered questions. I kept wondering if my life would have been better if I had returned to my family in Burundi instead of staying in Uganda. I prayed about it, and I think God gave me an answer to this question.

Moving with the chimps from Burundi to Kenya and then moving to Uganda is what ultimately saved my life. I know that so many men my age were killed in Burundi, especially the Hutu men. If I had stayed, I really think I would have been killed too. If the soldiers had come to our village and found me, I would have been taken away. My chances of survival were slim. So, the chimps saved me. And Debby saved me. They were the two things God used to save my life. I really do believe that.

Chapter 6

AN ISLAND OF SHELLS . . . AND CHIMPS

There was once a chimp living on an island in the middle of a lake in a country called Uganda. The chimp was named Ikuru, which meant "the happy one." Oh, it was such an ironic name, because when she was still only a baby, Ikuru had been stolen from the wild.

Poachers shot a female chimpanzee in the rain forest, and when they found the dead body of that chimp lying on the bloodied ground, they saw that she had been a mother with a little baby. The poachers knew a baby chimp was worth big money. People loved cute baby chimps, especially rich people, who were prepared to pay. But the terrified infant was holding onto her mother so tightly, she couldn't be captured. So, unspeakably, this tiny baby and her dead mother were thrown onto a roaring fire to force Ikuru to release her grip.

She was rescued from the poachers, but little Ikuru's suffering didn't end. A soldier tried to keep her as a pet. She became sick. Eventually, she was taken to a sanctuary and was nursed back to health by a group of caring humans. Ikuru grew to be a healthy adult. She loved to play and to groom her friends. She loved babies.

Finally, she had the chance to live a life worthy of her name—at the chimpanzee sanctuary on Ngamba Island.

Ngamba is a Luganda word that means "shells." Ngamba is one of hundreds of islands in Lake Victoria, which is Africa's largest lake. The lake is so big it's more like a freshwater sea. And just like a sea, Lake Victoria has its own underwater environment. There are many species of freshwater shellfish such as snails and other mollusks, and when the winds blow hard and the waters of the lake start to break and foam, millions of shells are washed onto the shores of islands like Ngamba.

Debby Cox first thought Ngamba meant "spiders." This would also be a good name, because there are a lot of spiders living on that island. Along with the spiders, Ngamba Island is home to snakes, monitor lizards, bats, thousands of birds, and millions of insects. There are also visitors from the lake like otters, hippos, and crocodiles, and even a few humans in their fishing boats. Today, however, more than any other animal, Ngamba Island is famous around the world for its chimpanzees.

As the nineties came to an end, the number of chimps at Entebbe Zoo was growing. Along with the chimps already living at the zoo, we were caring for another nine that had been living on an island called Isinga.

Isinga Island is in Lake Edward, in Queen Elizabeth National Park, about a six-hour drive west from Entebbe. The island was mostly rocks and bushes, with not many big trees. It was the wrong island for chimps. They were not in very good condition, and Debby was not happy with how they looked. I don't know how she did it, but she managed to convince the authorities to have them taken off that island and moved to Entebbe Zoo.

Removing those nine chimps from Isinga was a big challenge. They had to be sedated and then transported back to Entebbe. One of the male chimps on the island was called Maisko. He had been taken to

Isinga after he was confiscated from a Russian circus. Maisko was the boss of that island. As I said before, when you work with chimps, you must respect the boss, and so we had to respect Maisko. Debby and I had no dart gun, so we had to sedate each chimp by hand injection. When we went to inject Maisko, he charged me. He took hold of me by the arm and bit me. Even though it really hurt, I was able to push Maisko against a tree, and then Debby came over and injected him. Though chimps can inflict serious bites, I wasn't badly injured. And as we shall see later, Maisko and I would cross paths again.

There was another incident I remember from that time. Kidogo was an adventurous chimp. After Debby injected her with sedative, Kidogo climbed up into one of the few big trees on Isinga Island. Now, that tree was overhanging the lake and the water was deep. No matter what we said or did, Kidogo wouldn't come out of the tree. When the sedative made her fall asleep, Kidogo fell from the tree and splashed down into the lake. Even when they are conscious, chimps can't swim. And Kidogo was sedated.

Without a word, Debby ran and dived into the lake to save Kidogo.

Debby said she remembers being down deep in the water and looking up to see a chimp face staring back at her, in all the ripple and glare from the lake surface. She said it was something she will never forget—a chimpanzee face staring at her underwater.

It was windy and this made the water of the lake very rough. Even though she is a good swimmer, Debby needed help to get Kidogo back to shore. But I couldn't swim! I ran to the water's edge, and I reached out and was able to take hold of Kidogo's legs and pull her up and out of the lake. I felt happy when she and Debby were both back on dry land. I was so relieved that they were okay.

Things were getting crowded for the chimps of Entebbe Zoo. Along with the former residents of Isinga Island, we were receiving more orphaned chimps, particularly from Congo. We were running out of room.

Debby and some others from the Uganda Wildlife Conservation Education Center began visiting islands in Lake Victoria, searching for somewhere that might be suitable as a chimp sanctuary. After looking at many other islands, they visited Ngamba. Unlike Isinga, Ngamba Island seemed to offer everything that we required for a chimpanzee sanctuary. Debby was convinced. "As soon as I saw it," she told me, "I just knew it was perfect."

Ngamba Island is about a hundred acres in size. It's mostly covered in natural tropical rain forest. When I walked through the forests of Ngamba Island, it was like being in the heart of a great wilderness. The forests were so thick in some places, you could easily lose direction. Everywhere you looked there was life. There was greenery everywhere. It was beautiful.

And in those Ngamba Island rain forests there were trees that wild chimps in Uganda used for food. There were ants and termites and other insects, known to be popular food for the chimps. There were nesting birds to provide eggs and even small mammals that wild chimps hunted for meat. As well as the natural food of the forest, there were also exotic fruit trees that had been planted by some former human occupants of the island.

The northern tip of the island had been cleared of forest right down to the lakeshore, and we knew this would make a good place for landing material and people and eventually, the chimps themselves. The clearing would also be the perfect place to build the human facilities of the sanctuary.

Debby didn't waste any time. She wrote a proposal, which she sent to JGI. She also sent copies of her plans to other conservation organizations. She did this deliberately, because she didn't want the future of the sanctuary—and therefore the chimps—to depend on one organization. Chimps live for fifty years or longer, so Debby knew the sanctuary would need to be there for a long time. To survive into the future, the sanctuary management had to be strong and flexible. And, you know, more than twenty years later, Debby's vision continues to bear fruit. Ngamba Island is run by an entity called the Chimpanzee Sanctuary and Wildlife Conservation Trust (CSWCT), with founding trustees including the Jane Goodall Institute, Ugandan Wildlife Authority, the Born Free Foundation, Ugandan Wildlife Conservation Education Center (Entebbe Zoo), the Taronga Conservation Society Australia, and several other organizations.

Debby flew back to Australia to have a meeting with management of Taronga Zoo. They agreed to hold a fundraising dinner and invited all kinds of businesspeople to that dinner. On that one night, they raised half the money needed to buy Ngamba Island—more than $200,000 Australian. Debby also traveled to the UK and the US, where she was able to raise the remaining funds we needed to buy the island.

It was happening. The vision was becoming real.

Though I was excited about having a new home for the chimps, my first thoughts about the sanctuary were not good. The only way to get out to Ngamba Island was by boat. Lake Victoria is so big, and the water is very deep, and I couldn't swim. Even today, many years later, I can swim only a little bit. However, before I first went out to the island, I had a dream. In my dream God told me I wouldn't drown, so after that I was not so worried. I trusted Him, and I knew it would be okay.

Though it was officially an uninhabited island, Ngamba did have people living on it. For years, fishermen and squatters had been living on the island without paying any rent. And while the land had now been purchased for the chimp sanctuary, and anyone on the island was now a trespasser, no one was bullied or threatened to leave. Debby believed the process would be much fairer and more effective if we worked with the squatters instead of upsetting them. There was a local guy with a big boat who had started helping us take things over to the island. He spoke to the squatters and the fisherman, and he said they could have free use of his boat, but only if they agreed

Eddie could be very dangerous, but he was my friend. © *Barbara Hollweg*

Eddie used to escape a lot on Ngamba Island. © *Barbara Hollweg*

Working in the backyard of the halfway house, with Amizero. © *JGI Burundi*

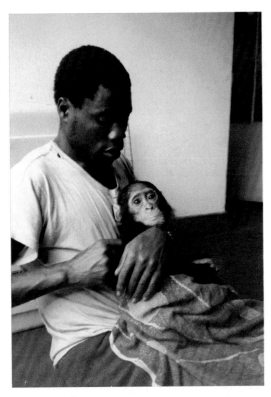

Caring for little Uruhara at the halfway house.
© *Aly Wood*

Poco at Sweetwaters in Kenya. Because of his cruel upbringing, he mostly
sits or stands on his back legs. © *Jenny Hobbs*

Working at Entebbe Zoo, with my brother Minani (on the left) and another Burundian refugee named Melanie. © *Debby Cox*

Doing our zookeeping certificate at Entebbe Zoo . . . but I couldn't read English. © *Debby Cox*

Aly with Nowera, Innocent, and Charles (standing) after fleeing Burundi. © *Aly Wood*

Debby Cox, Minani, and me along with Aly Wood on Ngamba Island. © *CSWCT*

The original chimp holding facility on Ngamba Island. © *Margaret Hawkins and CSWCT*

When the orphans arrived on Ngamba Island, we would stay with them constantly.
© *Barbara Hollweg*

Eating peanuts with Yiki and Pasa. © *Barbara Hollweg*

Debby is like my big sister. She taught me so much. © *Penny Tweedie*

With Pasa on Ngamba Island. © *Barbara Hollweg*

Returning Sunday the Boat Captain to Ngamba Island after he sailed off in a fishing boat. © *Susan Alancraig*

The chimps on Ngamba Island loved their morning porridge. © *Barbara Hollweg*

Mika was a very serious alpha male. Here he is with Kidogo. © *Barbara Hollweg*

With Nowera. My wife is strong and beautiful. © *Stany Nyandwi*

Inside my family house in Kawuku, Uganda. © *Jim Kurtz*

Because of the chimps, I have traveled the world.
© *Stany Nyandwi*

From their large exhibit at Taronga Zoo, the chimps have views over beautiful
Sydney Harbor in Australia. © *David Blissett*

With the team from Chimp Haven in Louisiana. © *Jim Kurtz*

My first meeting with sweet old Toddy, at the Center for Great Apes in Florida © *Jim Kurtz*

Speaking about chimps with villagers in the Hoima district of Uganda. © *Stany Nyandwi*

Speaking about chimps in Uganda. © *Stany Nyandwi*

Evening pant-hoots at Chimp Eden, in South Africa. © *Barbara Hollweg*

Thomas and young Tabu at Chimp Eden. © *Stany Nyandwi*

Claude escapes his enclosure at Chimp Eden. If you don't know the chimps, this is a very dangerous situation. © *Stany Nyandwi*

I have so much love and respect for Dr. Jane Goodall, here filming for the *Pant Hoot* documentary. © *Pant Hoot Film*

Stany on Ngamba Island © *Andy Nelson*

to leave the island without any fuss. He told them that, once he had finished his work with us, the offer would stop and he wouldn't be responsible for what happened to them after that. It was a good offer, because all the squatters left the island.

Despite the success of this process, Ngamba Island continued to receive illegal visits from fisherman and robbers. Sometimes, these visits were risky. Chimpanzees are naturally curious, and they soon came to know that boats carried food and wonderful things to play with. Chimpanzees and boats are not a good mix.

Even though they lived on only a small part of the island, the fisherman and squatters left a lot of mess behind, and we faced the challenge of clearing away all that mess and transporting it back to Entebbe. Debby offered the Entebbe Zoo staff positions out on Ngamba Island, but most of them didn't want to go. There were happy to keep their current jobs. Maybe some of them, like me, couldn't swim, and the thought of sailing across all that water filled them with fear.

With few staff willing to help, Debby began contacting people overseas, offering them volunteer opportunities to come and work with us. Many volunteers came! And every day we would travel out to Ngamba Island, and we'd spend the day clearing away rubbish and preparing the site for the first buildings. There was so much rubbish. There were tarpaulins and nets and fishing lines and plastic bottles and old clothes and batteries. I remember we sent boat after boat full of rubbish back to Entebbe.

Once the site had been cleared, Chris Kerr, an engineer from Taronga Zoo in Sydney, came over to Uganda to help us build some infrastructure. The first item to be built was a 1,300-foot-long fence to separate the chimpanzee forest from the cleared area where the administration facilities would be built. A mesh enclosure called the holding facility also needed to be built. The chimps would stay in this enclosure when they first arrived on the island. There were also plans to construct four brick buildings for storage, accommodation, and administration, as well as a timber observation platform for visitors to come and watch the chimps in the forest.

Chris Kerr looked like a real Aussie guy, with his wild hair, his wiry beard, and his floppy bush hat. Chris could build anything, and he still remembers the challenges of working on that island. There were tremendous rainstorms and an invasion of fire ants. He remembers the heat, the big swarms of lake flies, the lack of power, and the isolation. Building materials would be delivered late and sometimes they would be lost or stolen. And then, just to make his life even more challenging, during construction, Chris was told the BBC was coming to make a documentary about the sanctuary and they had a deadline to meet, which meant Chris had a deadline as well.

With water transport the only way to reach Ngamba Island, every roll of wire, every post, every brick, every sheet of iron, and every tool had to be carried across the lake by boat. At one point, Debby decided that the local boats—primitive-looking, canoe-based craft— were so small that moving everything across to the island was taking too long. She found a local entrepreneur who owned a large flat-top barge, and she hired him to transport large loads of materials and bigger groups of workers across to the island. A bigger boat would mean fewer trips, and fewer trips meant the work could be completed faster. It was a good plan.

Chris Kerr told me about the first time he saw the barge at Entebbe. "Yeah, it was a big barge all right. Now, I'm not a boat expert, but as soon as I saw it, I thought, that vessel just didn't seem right. She was flat-decked, almost square and sat on pontoons made of forty-four-gallon drums. At the rear there was a wheelhouse for the skipper to navigate the barge, and the thing was powered by twin outboard engines. No, it really didn't look like an open-water vessel to me."

The captain of the barge told Chris his vessel could transport for-ty-four tons of cargo at a time, but that the barge was mostly used for dance parties out on the lake. Chris kept looking at that barge and doing the math in his head. Something didn't add up. He was worried.

On the day of the first cargo shipment to Ngamba, I was working up at Entebbe Zoo. Chris and Debby arrived at the dock in Entebbe to find the usual captain of the barge had been replaced by another man. Then they were told that one of the two outboard motors wasn't working properly. They looked down at the barge and noticed it was already very low in the water, even though it wasn't fully loaded.

"We set off with six of our boys, four crew members, me and Debby," Chris told me later. "The four crew members were, somewhat unnervingly, all wearing life jackets. The rest of us weren't. Thankfully, there was no wind, and the lake was calm, because Lake Victoria can be really rough when the winds and storms blow. Anyway, about twenty minutes into the journey, the motor began to smoke and splutter and I noticed the barge had taken on quite a list in the water. Debby and I started throwing bags of stones overboard, but things weren't improving. In fact, they were getting worse! Debby decided it was time to call for help on her mobile phone. Remarkably, her call went through, and she was told the Ugandan Marines had been alerted, but that they wouldn't be on scene for at least forty minutes. We looked out across the listing deck of our barge and we just knew forty minutes would be too long. At this point some of us decided the best thing to do was abandon ship. Debby and I jumped off, but the African guys didn't follow, because none of them could swim. It was chaos on deck. There was plenty of hand waving and loud prophecies of an impending watery death. Now, I can swim a little, but Debby can swim a lot. I reckon she could have swum back to Entebbe if she wanted. And I really must thank her for swimming off to collect my precious Taronga Zoo hat, which had floated away when we jumped in the water. Thankfully for the boys and for me, some boats started to arrive and they rescued us, so no one was lost. But we did lose a lot of material. And losing that volume of stuff put us a week behind on an already tight schedule."

Meanwhile, from where I was working at the zoo, there was a view out over the lake towards Ngamba Island, and I saw what was happening to Chris and Debby and the others. I heard people screaming out in the water and I could see that the barge was sinking. I knew some of the people on board couldn't swim. I could see people clinging onto the barge or holding onto floating pieces of debris. I was relieved when I began to see those other boats coming along to rescue everyone.

Oh, and the other thing I remember was when Debby came ashore after she had been rescued. She was frustrated and worried . . . and very wet. Mostly, though, she was angry. She had made a bad choice. There were local boats that had been carrying cargo on the lake for thousands of years, and she thought a Western-style barge would be a better option. "Boy," she said to me, "did I learn the hard way."

From that time on, only the local boats were used to carry workers and materials over to Ngamba Island.

After those first facilities were built, we had to decide which chimps should be taken over to Ngamba Island. Debby and I worked together to identify the most suitable chimps for the first move. We decided the first group should consist of only females and some of the younger males. If there were any problems with the infrastructure, if the chimps broke out of the holding facility or through the fence separating the forest from the administration area, it would be safer if those chimps were females or younger individuals. The bigger, more aggressive adult males would be left for a future move, after we were satisfied with the facilities. We chose sixteen chimps to travel to the island in the first move.

This decision led to some interesting group dynamics. Without the adult males, a female chimp called Kidogo—she was the girl who fell into Lake Edward and was rescued by Debby—took over the role as alpha chimp. Kidogo was the biggest female chimp on Ngamba Island, and she remained boss of the group until the adult males arrived. She even walked like a boss chimp, full of swagger and authority! After the adult males arrived, Kidogo lost her authority and big Robbie from Entebbe Zoo took charge again.

Accompanying the first group of chimps was a staff team, led by myself and Gerald, who was the former head keeper at Entebbe Zoo. There were two other caregivers as well. Four caregivers were not enough. Tourism had always been a part of planning for the Ngamba Island Sanctuary. Debby came up with an idea. She arranged for thirty volunteers to come from overseas to help with the early work. There were no accommodation buildings to begin with, so these volunteers stayed in tents. For a while, the cleared area of Ngamba Island looked like a tent city.

As part of our tourist plans, we thought about allowing tourists to take forest walks with the chimps. You can do this at places like Kibale Forest in western Uganda or at Gombe in Tanzania. So, for three months we would walk every day in the forest with the chimps. At first, they were afraid, but gradually they got used to the sights and sounds and smells of the forest. We had guides from Kibale come to observe and give us some advice. Eventually, we gave up the plans to do forest walks in this way. Because most of the Ngamba Island chimps were orphans, they had been imprinted on human beings from when they were little. They had lost all fear of humans, and they would jump on visitors and try to play with them. This is OK while a chimp is little—in fact, until recently, visitors to Ngamba Island could do short forest walks with the youngsters—but once a chimp becomes an adult, they can be too dangerous, especially if they are used to interacting with people. In the wild, at places like Kibale or Gombe, you can walk in the forest with adult chimps, but these chimps are truly wild, and even though they are used to visitors, they have not been imprinted on people. There is no physical contact between chimps and humans in these places.

We couldn't guarantee this at Ngamba Island, so we had to abandon the idea.

Perhaps the most exciting day on Ngamba Island was when we first opened the sliding metal door and let the chimps walk up the raceway from the holding facility towards the forest. The chimps came out of the holding enclosure and began to move up the raceway. They began pant-hooting, following closely together and touching each other for reassurance. When they came to the end of the raceway, they stopped and fell silent. The first chimps stopped at the opening of the raceway and peered out into the forest. Then they looked up to me and the other caregivers for reassurance. Eventually, led by Kidogo, they began moving out of the raceway and into their new forest home.

Those early days of exploration were exciting for us as well as for the chimps. Each morning, there would be so much running and jumping and swinging and noise. It wasn't long before the chimps lost their fear, and then they really enjoyed going into the forest. As well as the natural food that grew there, were some garden plants that the fishermen and squatters had planted—figs, passion fruit, and ginger. When the chimps found these trees, they ate and ate and ate. Some of them were so greedy it looked like they were pregnant! Then, each evening, the chimps would come back down the raceways and into the holding facility again. They didn't like spending the night in the forest.

While development of the sanctuary continued, some of the caregivers would stay out on the island over several days. During the day we worked to feed and clean and care for the chimps, and at night we slept in the newly built staff quarters. We stayed there to watch over the gear and assist if someone came calling. Then one evening . . . someone did.

I remember hearing something moving around outside. I thought it was hippos, because they sometimes came out of the lake to eat the grass on the cleared part of the island. The sounds came closer and closer. Gerald and I were alone in the staff quarters. When I heard voices close to where we were sleeping, I realized the visitors were not hippos but humans. The intruders made their way directly towards the storage areas where we kept valuable items like tools, a generator, and an outboard boat motor. It was obvious they were coming to steal our gear. And I also believe—even to this day—that they were people familiar with our setup, because they knew exactly where to go, even though it was very dark.

Gerald and I were alone in the bedroom. There were robbers now sneaking around, and we didn't know how many there were. We didn't know if they were armed. I felt like I had to do something. So, I got up and moved quietly around the building until I found an empty jerry can. I picked the can up, and I threw it down hard on the ground, so it made a lot of noise. The robbers must have been startled, because they began firing shots. This scared the chimps in the holding facility, and they began to make loud fear screams. There was so much noise the robbers must have been scared, because they ran off.

Maybe it was foolish of me—we knew they had guns—but I started chasing them down towards the lakeshore, where they had a boat waiting. I was unarmed. I shouted, "Hey you, come back here!"

"Yes, we'll be coming back," they shouted in reply. "And when we do we're going to get your motor and your generator and we'll get you as well!" Then they sped away from the island in their boat.

A little while later, Gerald and I heard another boat coming towards us. We were sure that the robbers were coming back. Maybe there were more of them this time. Maybe they had even more guns. So, Gerald and I ran out of the building and hid in some bushes where the robbers wouldn't be able to find us.

A voice called out. The voice seemed to know our names. How had the robbers worked out our names?

But as the boat came closer, we realized it was some friends from a neighboring island. They had heard the shots and the screams and had come over to make sure we were okay.

At first light, exhausted after a night without sleep, I went up to the holding facility to check on the chimps. Though none had been hurt, the holding facility was in terrible condition. Maybe you've experienced what terror can do to your stomach. Well, it's the same for chimps, and that enclosure was absolutely plastered in chimp diarrhea.

The police came out to investigate the attempted robbery. Gerald and I were asked to give statements. After the police completed their investigation, it was decided that an armed officer would be stationed on Ngamba Island at all times. They would provide security for staff, visitors, and the chimps.

As the chimps became used to life on the island, we realized some modification and reconstruction was going to be needed. Major changes had to be made to the holding facility. This enclosure was never planned to be permanent. We thought at first that the chimps would stay in that enclosure for only a few nights, until they were used to the forest. We thought they would start to sleep out in the forest, building nests in the trees like wild chimps do. But it didn't happen that way. The chimps didn't like being out in the forest at night. They wanted to come back to the safety of the holding facility. So, before we brought in more chimps, especially the big males, we had to make the holding facility enclosure larger and more secure. In the same way, the two raceways that linked the holding facility to the forest were only temporary. They were enclosed using lightweight wire mesh, and an adult chimp—especially a big male—would be able to break through this wire without too much effort. So, the wire was replaced with more robust mesh and metal bars. And still, more than twenty years later, the chimps of Ngamba Island return to their night enclosure in the holding facility each evening. Here they have hammocks and suitable materials to make sleeping nests. Few, if any, chimps ever sleep out in the forests.

Also, the electric fence separating the forest from the administration area wasn't big or strong enough. The chimps started breaking through the fence so they could explore the wonders of the administration area. This was annoying when the females and juveniles did it—chimpanzees are not known for their neatness—but it would become potentially dangerous once the adult males arrived. So, a second, much taller and more robust fence was built. Once these construction jobs were finished, the adult males were brought over to the island.

Even after the larger fence was built, some of the chimps still managed to find their way through. As you read in the prologue to this story, Eddie was one of the chimps who used to break out of the forest. And he wasn't alone. Still, the incidence of escapes was significantly reduced after the second fence was built.

So, there have been a lot of changes and improvements on Ngamba Island since those days of the late nineties. People visit Ngamba Island today, and they see all the wonderful infrastructure and facilities and think it was always like this. But I can tell you, it was very different at the start. So much hard work had been done.

While the modifications improved the overall safety for both chimps and humans on Ngamba Island, there were, and still are dangers present. The island remains largely wild, as is the surrounding lake. All manner of wild creatures live on, or around the island. And, the intelligent chimps soon came to know the difference between the animals that posed a risk and those that did not.

The hippos and crocodiles that live in Lake Victoria would sometimes visit the island. The chimps would avoid them, because they knew these animals were dangerous. There are many spiders and a lot of snakes in the forest. Some of the snakes are venomous. And, while they would mostly stay away from snakes, I knew some of the chimps were able to tell the difference between the venomous and nonvenemous species. I have seen chimps harass and hit a python to chase it away, but I have never seen a chimp do that to a venomous snake like a cobra.

And then, sometimes, it was the wildlife on Ngamba Island that was in danger. The chimps loved eating insects. Termites were their favorite. They also killed other, larger animals, but when they did this, they didn't seem to eat the animals. Chimps in the wild do kill animals for food. Some wild chimp communities are among the most efficient hunters in Africa. On Ngamba Island, however, they seemed to kill other animals more for fun. Natasha was one chimp that hunted a lot. She liked to catch and kill the mongooses that lived on the island. But she never ate them. Once they were dead, I used to see her carrying them around on her back, like they were her toys.

Today, more than twenty years after its creation, the Ngamba Island Chimpanzee Sanctuary remains one of Africa's most celebrated great ape facilities. The island receives thousands of visitors each year, and it funds local community and conservation programs. I was fortunate enough to be there at the beginning. And I saw how much work people like Debby Cox and Chris Kerr put into that sanctuary. I think it was Debby and Chris and the other volunteers coming from Australia that helped make Ngamba Island such a big success. I used to say, "Ngamba Island was the sanctuary that came from far far away."

Of course, Debby deserves a lot of credit for how good that sanctuary became. She never asked us to do a job she wouldn't do herself. People understood her, and people listened to her. She listened to other people as well and she trusted them. I always felt like Debby trusted me.

It was during my time on Ngamba Island that my reputation began to really grow. Working at the sanctuary was never really just a job—it became an important part of my life. I was able to refine the skills and knowledge I gained during my time at the halfway house, Nairobi Orphanage, Sweetwaters, and Entebbe Zoo. This was the time when I learned the most about chimps—and myself.

Chapter 7

REPUTATION, INTEGRATION, AND CHIMP TALES

For a long time, people have called Doctor Jane Goodall a "chimp whisperer." Her work at Gombe opened the eyes of scientists and the wider world to the true nature of chimpanzees. It was Jane who first opened a window to let us see what our closest biological relations were really like. Because of this, she is celebrated and loved. So, you can imagine how special it felt when someone like Doctor Jane started referring to me as a "chimp whisperer." That was such an honor.

I learned more about chimps while working on Ngamba Island than at any other time in my life. I had the chance to take what I'd been taught by people like Dean, Susanne, and Debby, and I started to work with chimps in new ways. Still, no matter what new things I tried, the basic principles never changed.

If you want to work successfully with chimps, then you need to know them—as a species, as a community, and as individuals. Everything they do has some sort of meaning. So, you must make the time to stop and watch and listen. I have done this over many years. Here are just some of the things I've seen and learned.

*

I have learned that there can be subtle differences between various chimp calls, and how, when combined with gestures, the same calls can mean different things. Sometimes people have asked me if I can really speak chimp. In other words, because chimps react to me as they would to another chimp, do they really consider me to be one of them?

The simple answer is no. The chimps know I'm different from them, but they do still understand me. How do I explain?

It's like when I communicate with someone who has English as their first language. Yes, I can speak English, and I can write in English, but not so great. I make many mistakes. Still, people who speak English very well can understand me. And I believe it's the same with the chimps. I sound like a chimp, they understand me, and we can communicate with each other, but they know I'm not one of them.

Many people, including scientists and researchers, have spoken about how strong chimpanzees are. The most common theory is that an adult male chimp is between five and ten times stronger than an adult human. I don't know if there is an exact number. And I'm not a scientist, but I have seen for myself just how strong a chimp can be.

Many chimp sanctuaries or zoos have raceways, which are long tunnels covered by wire or metal mesh that join one enclosure or exhibit to another. There are raceways like this at Ngamba Island, connecting the holding facility to the forest. At the end of these race-ways there are metal doors that slide open and close. Sometimes a chimp doesn't want to be moved, so they will block these doorways to stop them being opened or closed by the humans. I know that one male chimp can hold open a sliding door like that, even when there are three or more human men pushing to try and get that same door closed. I have seen this myself—one chimp keeping a door open because he was stronger than a group of humans.

And it isn't just strength. Chimpanzees are intelligent and emotional. Each chimp is as individual as each human. And, just like with humans, no two chimp relationships are exactly alike. You need to know each chimp personally and their relationship with all the other individuals they associate with. If you don't do this, you cannot understand how their community works. I can give you a good example.

Chimps practice what we would describe in human society as "politics." And sometimes, when I see the way human beings do politics, I think we copied from chimpanzees! When a chimp group is selecting a leader, this can sometimes take a long time. There will be much fighting and testing between the leadership rivals. Alliances are made and then they are broken. Even if he—and alpha chimps are almost always adult males—is beaten seriously in a fight, a good potential leader will get himself back up and try again.

But chimp power struggles are not just about being physically strong. A good leader must be smart as well as tough. Robbie, who was the alpha male at Entebbe Zoo, remained the boss after the group was moved to Ngamba Island. However, there was another chimp, a big and powerful boy who I've spoken about before, by the name of Eddie. He started fighting with Robbie, and he used a lot of violence to make it to the top position. But Eddie could never stay as the leader of that group for long. He was strong, but he wasn't very smart. He was the boss for only three months. Now, there was another male called Mika. I had worked with Mika since he was young. He was a very smart chimp. By the time all the chimps were moved to Ngamba Island, Mika was grown up and he was able to challenge for the alpha role. Mika had the support of many other chimps. He would display often. He would watch over the group. He was very serious about being an adult male. We used to call him the "City Boy," because he was so streetwise. He was always aware of the politics. The other chimps, including Eddie, knew how smart he was, and Mika took over the role of Ngamba Island boss without any major violence.

You know, I like to think that the two of them sat down one day and Mika said to Eddie, "Let me be the boss, and you can be my number two."

Eddie agreed. And Mika stayed as the alpha male on Ngamba Island for more than ten years.

While they almost never compete for the alpha position, females have their own hierarchy and play an important role in choosing the overall group leader. So, the females will watch the males and see who is strong. But they are smart too, and if they don't like one of the males and they see that male being beaten up, they will join in the fight to make sure he is kept down. And a female may also jump in to help her favorite male when he is in a fight. The really smart females make friends with the alpha male, because this improves their position in the hierarchy. Sometimes you will even see females—and the lower-ranking males—supporting more than one leadership rival. They do this so that no matter who wins the top job, their status will rise as well. Having the boss as your friend is a good thing when you are a chimpanzee!

So, why is it important to be the alpha male chimp? Well, there are many reasons. First, the boss gets the choice of the females. Second, he gets the choice of the best food. And third, he gets the respect of everyone in the group. The other chimps will show him respect. They will greet him, sometimes you will even see them lying down in his presence. Sometimes, I see a boss chimp just look at another chimp and that can be enough to stop that other chimp from doing something wrong. A look can be enough to send another chimp running away. A look or a gesture and the other chimps will go running to the boss to pay their respects. In chimp society, the alpha is the king.

These things are very good, but the alpha has responsibilities too. If there is any fighting, or the group is under threat, then he must be involved. He may let lower-ranking chimps sort out their own fights, but if the fights are serious and threaten the group, or if there is an outside threat, the alpha male is expected to lead from the front.

What's most important for me is the way chimps do power and politics directly impacts the way caregivers manage that group. As caregivers, we must understand the rank structure, and we have to respect it. We also have to hold some kind of place in the structure. So, when I was caring for the Ngamba Island group, then I had to be the alpha in certain situations. For instance, when a chimp escaped from the forest and into the administration area, they needed to know that was my area, and in that place, I was the boss. This was how I got Eddie to go back into the holding facility without sedating him when he escaped at Ngamba Island—you know, this was the incident that David wrote about in the prologue to this book. In the administration area, I was the boss, and once Eddie understood that . . . no problem.

There are other reasons chimp caregivers must understand group politics and dynamics. You must know each chimp and their place in the group. When you know who the boss is and who has low rank, then you can help manage the group properly. For example, if a low-ranking individual is missing out on food, you need to be aware of this and work within the dynamics of the group to make sure that chimp gets enough nutrition without upsetting the group structure.

I never give up on chimps, no matter what. I will develop a relationship with every single one of them. I have to tell them when they are wrong or being naughty. I have to submit to them when they are right, especially when they are the boss. Too many chimp keepers and caregivers don't understand this. They want to get their own way all the time, but this doesn't work with chimps. Sometimes, I even see keepers who are scared of chimps, especially the big males. The chimps know when someone is afraid of them, and they will take advantage of the situation. Me, I respect them, and I want them to respect me. This means that sometimes I have to step back from a situation and give the chimps time and space to work things out for themselves. This might be a problem for me, but I have to do it. Only when caregivers and chimps work together can management really be successful.

Successful integration is one of the toughest parts of being a chimp caregiver. Integration means safely introducing a new chimp or chimps into an established community. It's something we did a lot at Ngamba Island, because orphaned chimps continued to be confiscated and brought to the sanctuary. Chimps may seriously injure or kill strangers, so integration must be done with all care and understanding. Caregivers must recognize and respect each individual's personality, rank, or status.

When a new chimp—and they were often infants—first came to Ngamba Island, a caregiver would stay with that infant for twenty-four hours a day, just like its mother would have. Building this rapport was very important for both the chimp and the human. It provided comfort and supported the well-being of each young chimp, and it allowed caregivers to begin to learn and understand the individual personality of that chimp. Remember, the key to success means knowing every individual. If you don't know each chimp, the integration process will be much harder and you may not be successful.

On Ngamba Island, newly arrived orphans were never blood relatives of the resident chimps. This made integration harder. So, we used to introduce the new orphans to certain adult females who were gentle and had good mothering instincts. At Ngamba Island, we would often use two chimps named Connie and Becky for this purpose. We used to call them our "welcoming committee."

Then Debby and I came up with a new idea! We would take the juveniles and the welcoming females into a small area of the forest that was separated from the main chimp forest. Working in this environment gave everyone more room to move, and they could get to know each other without feeling trapped. Once the juveniles formed a bond with the welcoming females, the task of integrating them into the larger group became easier, because the little ones now had friends within the group. After they were integrated with the females and lower-ranking males, we began to integrate new chimps with the big males, especially the alpha.

However, even the best integration processes involved risks. Integrations can be stressful, and they sometimes come with consequences—for both humans and chimps. If you integrate a chimp into a new group and that chimp is killed, there can be big problems for the sanctuary, so you must take your time and you must get it right.

Sometimes, no matter how well you are prepared, the chimps will fight and attack a new member. It can be scary to watch, but you have to step back and let them sort out things for themselves. Fights are fairly common, and they are one way that chimps work out their rank. Sometimes, though, you would have to step in if it meant saving a chimp from being killed or seriously injured.

On Ngamba Island, I developed a technique to allow me to step in when a situation became dangerous. It's what I called the jackpot approach. If a certain chimp was making trouble during integration, then it was important to know what they liked and what excited them. And just like people, different chimps were excited by different things. I would introduce these exciting things as diversions during integration. Mika was one chimp with whom the jackpot approach worked very well.

Mika loved his morning porridge. All the chimps were given porridge each morning, which was made by mixing millet flour with boiling water and adding a little sugar. When I did an integration with Mika, I would give him extra porridge with lots of sugar in it. Mika loved sweet porridge, so this was his jackpot. I used to say to him, "Mika, here is your jackpot . . . this is just for you." He would get excited when I did this. And that was good, because when I was working with the alpha, I wanted to make sure he was feeling great. We would shake hands, and I would groom Mika through the bars. I would look for signs of contentment. When Mika started to roll on the floor and sigh, I knew he was content and happy and this would improve our chances of successfully integrating a new chimp.

Integration with an alpha male like Mika would be completed over stages. Each day, while Mika was eating his jackpot I would introduce the new chimp, sometimes through mesh, or sometimes face-to-face. I would do this for a short time, and then the new chimp would be separated again. The process would then be repeated, for a little longer each day, until the new chimp was accepted. Though the process could be slow and sometimes stressful, integration with the alpha male was essential. Any chimps not accepted by the alpha are destined to live separated from the group for their own safety.

As a rule, females are easier to integrate into a community than males. In the wild, females will sometimes move between communities, often when they are sexually receptive—and you can tell when a female chimp is in season by the large pink swelling she develops around her butt. When the females are pink, this becomes like a big jackpot for the males. The alpha male can be especially eager to have her in his company, and this means he will accept her very quickly. Integrating a female in season is much easier than any other situation.

Sometimes people will say to me that they don't like chimps. They say that chimps are dangerous and violent. Chimps are known to be cannibals, and people obviously don't like that. But I know that many people don't like snakes, but some people do. Some people don't like dogs, and other people do. To me, it is the same. And I think with chimps, when people see things like violence, perhaps it is the poor reflections of human nature they don't like and not just the chimps themselves.

Chimps are so much like us. Each individual is as different from the next as each individual human. There are shy and friendly chimps. Some chimps are very smart, and some are a bit slow. Some are greedy, and some are fussy eaters. I have worked with creative chimps and some that are plain and dull. Some chimps are ambitious for higher rank, while others are happy staying where they are.

And just like humans, chimps are sometimes known for their bad habits. Many of these habits are developed in captive situations, but not always. Some chimps are adept at throwing, especially males when they display. They do this in the wild. A displaying chimp might throw sticks or rocks to intimidate rivals. In human care, they throw these things too, plus anything else they can find. Sometimes they throw their own shit.

As I noted earlier, Eddie was a thrower. Eddie could throw far, he had a strong arm, but his aim was never very good. At Ngamba Island, he collected stones that he would pick up and use during his displays. It wasn't just for the other chimps—Eddie liked to throw stones at the human visitors. We would see him start to puff up his hair and pant-hoot, and we would warn the visitors to be careful. Then, we would watch him closely. As he started to display, if he reached down to the ground, we knew he was collecting one of his stones. He would run straight towards the fence and toss a stone, and we would all duck for cover. Thankfully, Eddie's throws usually missed. However, to improve his chances of a hit, he learned to collect a lot of stones and he threw them all at the same time. Sometimes you would see his stones flying in all directions. And I can tell you, if one of them hit you, oh, it really hurt! Thankfully, not all the males throw things.

Grooming, however, is something almost all chimps love. As well as removing parasites and dead skin, grooming is a very important part of building bonds among individuals. It is a social activity. Therefore, when I work closely with chimps, I will groom them. On Ngamba Island I would groom the younger chimps in person, unimpeded by a barrier, but with the adults I had to groom them through the mesh or the bars of an enclosure. And it was important that I let the chimps groom me.

Now, with some chimps, grooming became a painful experience. Some of them would groom me really hard. If they found a pimple or a spot on me, they would start to squeeze it. And they would keep on squeezing . . . squeezing . . . squeezing until it bled. That really hurt!

As well as learning many things about chimps, living with them on Ngamba Island for so long meant I collected a lot of wonderful stories. And, when I think about Ngamba Island stories, Sunday is the chimp who usually comes first to my mind.

Sunday was never an average chimp. Unlike gorillas and orangutans, chimpanzees show only minor sexual dimorphism—this word means the difference in size and appearance between the males and females. However, adult male chimps are slightly larger and more muscular than females. The males tend to have broader backs and shoulders, they often have longer hair, and their faces have bolder features. Sunday was a male chimp, but he really didn't look like one.

Sunday had come from a Russian circus. Like many performing chimps, he had been castrated when he was young. So, even after he had fully matured, Sunday looked more like a female than a male. And there was more. Sunday wasn't a normal-shaped chimp—he was tall and thin—and that was good because of how much he loved the water. In fact, Sunday was one of the reasons a 110-yard exclusion zone was put in place on the waters surrounding Ngamba Island.

Chimps can't swim, and many are afraid of the water. On Ngamba Island, though, some of the chimps became used to the water and would wade out a little way into the lake. But Sunday would go in as deep as he could, and because he was so tall, that meant very deep indeed!

We used to go to the fishing villages on neighboring islands and tell them not to come too close to Ngamba Island, because the chimps were dangerous. But some of the fishermen were curious about the chimps. And because so many other parts of Lake Victoria had been overfished, the protected water around the island became one of the few places where there were still many fish to catch. This attracted the boats as well.

The fishing boats were wonderful craft to our curious chimpanzees. They came loaded with food and clothes and tools and all sorts of toys to play with. Chimps know what stealing is, but that doesn't stop most of them from doing it. Sunday and some of his friends began ambushing any fishing boats that came too close to Ngamba Island—and then stealing everything they could find.

I would know when a boat had come too close to shore because I would hear the sound of chimps pant-hooting across the island. It was like they were saying, "Hey everyone, come here and look what we found."

When I heard this, I would follow the sound of the calls, and I would find the chimps with food and pots and nets and gumboots and clothes spread all over the place. In the water and on the shore, there would be stuff everywhere. Oh, they would make such a big mess.

But then, there was the time when a boat came too close to the island, and Sunday was waiting, and even though the boat was still out in the deeper water, Sunday waded out and jumped aboard. Of course, the fishermen got scared and all jumped out. Then some of the other chimps followed Sunday into the water, they took hold of the boat and started rocking it up and down to make waves. The boat began to float away. And Sunday was still onboard.

Fishermen across the north of Lake Victoria began reporting a boat sailing along with, apparently, no one onboard. But then, when they looked closer, they saw a black shape moving about in the boat, and that was Sunday.

Some of us jumped into one of the sanctuary boats and went to search for Sunday. Eventually, we found him sailing along, his hands reached out and holding onto the sides of the yellow boat like he'd been sailing all his life. We tied a rope to the front of "his" vessel, and we towed Sunday back to Ngamba Island.

Once on shore, Sunday returned to the forest, and the boat was returned to the fishermen. It was a big mess. There was shit everywhere in that boat. So much shit. And you know, to this day people don't believe the story, but I know that one of the volunteers took a photograph of Sunday while he was out in that boat, so there's evidence to prove it's true. And after that day, Sunday was nicknamed the "Boat Captain."

Like Sunday, Maisko had come from the Russian circus and, like Sunday, he'd been castrated. But unlike Sunday, Maisko was often an aggressive chimp. I first met him on Isinga Island when he bit me while Debby and I were trying to sedate him. On Ngamba Island, Maisko became a regular source of fear among the other chimps—and some of the humans.

But one day Maisko fell ill. Very ill.

I still remember the afternoon. The chimps were coming out of the forest for feeding. Straightaway, I could see Maisko had a problem. He was not walking normally. He was sliding along the ground on his backside. I wondered if he'd been attacked, or if he'd been bitten by a snake. Did he have some kind of sickness? It was hard to know.

The chimps were returned to their night enclosures in the holding facility. Maisko was isolated. He appeared to be paralyzed in both his arms and his legs. I could see he was distressed. As the hours passed, his condition became worse. We made a call to a vet at Entebbe Zoo. Maisko was sedated and various medications were tried. X-rays were taken. Blood samples were sent for analysis, in both Uganda and in the United States. As the results came in, the mystery grew. There was no definitive diagnosis. The only thing we knew for sure was Maisko was getting worse.

With Maisko lying motionless in his own excrement, his situation seemed hopeless. He had no energy and limited movement. He refused food. He failed to respond to any medication. The experts ran out of answers. They saw no hope of recovery. People started saying he should be put to sleep.

I didn't agree with this decision. I could see when I looked into his eyes—he was asking me to help him. And I wanted to help my friend. Accompanied by a courageous and dedicated nurse named Diana, who had been visiting Ngamba Island from the Caribbean, I did what I could to save Maisko's life. Diana and I went into the cage where Maisko lay. We began to speak to him, stroking and massaging his limbs. We had to be careful. Even though Maisko's limbs were paralyzed, his head and body were not. With his jaws and enormous canine teeth, a bite could be serious. Maisko had a reputation, and he'd bitten me before. Still, we continued to stay with Maisko, talking to him and massaging him. Over time, we came to understand where he had pain and where there was no feeling. Despite our efforts, he didn't improve.

I wasn't going to give up. I stayed with Maisko for hours. Days. I kept giving him massages. I made a bed of hay, turning Maisko regularly so he didn't develop pressure sores. I cleaned up his feces and urine. I spoke gentle words of encouragement. I even prayed.

Miracle is a word some people don't like. Many people don't believe in miracles, but I do. Because sometimes, when science and logic fall short, love and faith and hope do rise. Maisko rose as well. He began to move his limbs. And he began to eat. Soon he was able to reach for the bars of the night enclosure. He could climb up to his favorite hammock. He began to eat more and regain his body condition. He went on to make a complete recovery. The condition never returned. And no one really knows what the cause was, or how Maisko managed to recover. But, I think God knows.

A lot of the chimps on Ngamba Island earned themselves a nickname. Mawa certainly earned his. We called him the "Escape Artist," because Mawa liked to jump the fences separating the forest from the administration area. He didn't seem worried about the sting of the electric wires. He had a good reason.

Mawa came from the Congo. He was the pet of a rich man who used to walk him on the beach and take him to clubs and bars at night. I don't know if Mawa was ever given alcohol, but he didn't behave like most chimps. He was crazy! And he was always being rejected by the other chimps. He didn't mix well in the group, and he refused to submit to the alpha male, Mika. He didn't participate in community life—things like grooming—and he used to get beaten up a lot. So, I think he decided it was less painful to get an electric shock if it meant he could escape the other chimps.

A round-bodied fellow, with dark wise eyes, Mawa was moved to a permanent home in the holding facility, with some other escape artists. The decision to keep Mawa this way wasn't easy, but it was a matter of safety. Mawa decided that when he was out, even though he was in the human area, he would be the boss. He would scare people. And once a chimp knows people are scared of him, he grows bold. This can make him dangerous to work with. And Mawa also learned that when the people ran away, he could get all the food and toys he wanted. He would go into the administration area and find things to play with. You know, one day we found him with a mop—and he was using it to clean the floor!

Mawa even learned how to foil attempts at sedation. He would go to the hay store when he saw the vet coming with the dart gun. He would gather up big bundles of hay and hold them in front of his body like a shield. When a dart was fired from the gun, it couldn't hit his body.

Mawa's escapes always meant extra work for us. We had to relocate him back to the forest side of the fence and clean up the mess he left behind in the administration area. But there was one escape that became very serious. A scary day!

It was about two in the afternoon. Mawa was walking around the staff quarters, eating food and making mischief. He walked into one of the rooms where a police officer was staying. As I mentioned, after robbers had visited the island, we had a police officer staying with us for security. The officer was not in his room, but his gun was! He had an AK-47 rifle, which was supposed to be locked away in a secure cabinet. But this officer kept his gun under his mattress. Mawa threw the mattress off the bed, and he found the gun. He started playing with it. He put the gun down on the floor and started jumping on it. He picked it up and pressed the trigger. It is a big blessing that there was no round in the chamber and the gun didn't fire. But there was a magazine attached. If Mawa had worked out how to cock the gun and he pressed the trigger again, it would have fired. At one point, Mawa picked the rifle up like a caveman's club and started bashing it down on the ground.

It was then that the policeman turned to me and said, "How am I ever going to put this into a report?"

Because Mawa had escaped so many times, he knew our tricks to get him back to the holding facility. He knew we would use food as a lure. He knew how to deflect the vet's tranquilizer darts. He wouldn't submit to anyone. Mawa thought he was pretty smart, and now he had his hands on an automatic weapon loaded with ammunition. He had only to slide one lever on that gun and it would have been ready to fire.

I had to come up with a new plan. It had to be fast, and it had to be something Mawa was not expecting. I decided to go fill a cup with boiling water, and then I came back and walked straight towards Mawa . . . slowly . . . slowly. He stared at me, wondering what I was doing. I waited until he put down the gun, then I threw the hot water towards him. This gave Mawa a shock and he ran away. I picked up the gun and gave it back to the policeman. I don't know which one of us was more relieved.

Following this dangerous excursion, Mawa's time in the forests of Ngamba Island came to an end. He was moved permanently to the holding facility. I worked with the other caregivers to come up with new enrichment activities to keep his mind active and occupied.

Still, life in the holding facility didn't completely foil Mawa's mischief. I remember an American lady staying on the island. She loved

Mawa and kept walking up to the holding facility to talk to him. And Mawa knew she loved him! He would make play faces at her, pretending to be her friend. I knew Mawa was being sneaky. I knew he was up to no good! I kept asking the lady to stay back from the bars, but she wouldn't listen. She kept going closer and closer.

Then I heard a lady screaming. Oh, it was a terrible scream!

I ran up to the holding facility and found Mawa swaggering around his enclosure with that lady's bra around his neck and her handbag over his arm. When she had gone too close to the bars, Mawa reached out, snatched the strap of her bra and pulled it right off. He'd also stolen her handbag.

As we watched, Mawa opened the handbag to see what was inside. He found lipstick and a pen and the lady's period pads. The lady was embarrassed and angry, but it could have been much worse. If Mawa had caught her by the arm and not just her bra strap, she could have been bitten and seriously hurt.

Sometimes, even the staff were caught out by Mawa's trickery. One time I saw him steal the spectacles off one of the caregivers. And Mawa knew exactly what to do with them. He put them on his face and then he started walking around his enclosure. The spectacles made his eyes look as big as tennis balls, and he was staring at everything . . . staring . . . staring . . . like his eyes had gone funny. Mawa always kept us on our toes.

One of the things that keeps a male chimp on his toes is the chance to mate. When he's in the mood for sex, everything about a male chimp seems to be hard and erect—his behavior, his hair, his posture, and especially his penis. He might run around and display, slapping and hitting and chasing away the other males to show the females how big and strong he is. He might grab hold of a branch or part of a tree and violently shake it. Sometimes he will stand straight and tall on his hind legs and sway from side to side, his penis sticking out like a flagpole. Or he might stay on all fours, his arms held straight while he bounces up and down on the knuckles of his hands. Eventually, if he is persuasive enough—and especially if he is the alpha male—a female will concede and present herself to him for mating.

But, in most chimp groups there are other males beside the alpha, and they want a girlfriend too. And, you know, sometimes the girls don't fancy the boss! They might mate with him out of obligation, but they really want to have a different boyfriend. On Ngamba Island, we used to watch the various love interests of the chimps. It was like watching a soap opera on TV.

Katie was one of Ngamba Island's most popular girls. She really loved the boys. She was good at luring them away for sex. And all the boys wanted her. There could be three or four females in season, but if one of them was Katie, she would be the one the males all wanted. And she loved the attention. So, she would stand close to the alpha to show him that she was his girl, but then, at the same time she would start giving little signals to another one of her boyfriends. Then, while the alpha male looked the other way, Katie would disappear into the bushes with her other boyfriend. If she was ever caught with another male, Katie would make friends with the boss quickly so that he would take out his anger on the other male instead of her!

Aside from the alpha male, Katie had another favorite boyfriend. Tumbo was a big chimp, with silver hair on his head and down his back and legs. Though he never was the alpha male, he was a handsome boy. Even the human females visiting Ngamba Island would tell me how handsome Tumbo was! Katie certainly loved Tumbo. She would sit together with him a lot. She groomed him closely. And there was one special thing she did for him. Sometimes she would bend down, put her head between his thighs and suck on his penis. I'd never seen a chimp do that before I met Katie. And I think Tumbo liked this very much!

As well as being popular with the boys, Katie was also very smart. I remember a time when one of the caregivers dropped a key that landed in a place where the chimps could reach. Katie found the key. Then, I watched in amazement as she walked up to one of the padlocks of the holding facility, pushed the key into the padlock, and made it pop open. I had to rush up to the enclosure and close the padlock again before she escaped. You see, Katie had watched the humans using keys to open padlocks, and she learned exactly how to do it herself. Now she had a key that would open padlocks everywhere! This was a problem.

Katie understood the concept of "value." She knew that a padlock key was important to us. She was prepared to blackmail us to get it back! And Katie drove a very hard bargain! I kept saying to her, "Katie, bring me the key." But she refused to do it until I offered her a really good bribe—like a banana. If the payment wasn't to her satisfaction, Katie wouldn't cooperate.

Oh, there are so many other tales to share. There was Okech, one of Ngamba Island's more stubborn chimps, who became an expert at fishing for food. Okech would make himself a tool out of a stick or branch and he'd reach under the wire to collect any morsel of food that fell on the wrong side of the electric fences. He was fussy too. In the wet season, the food would get muddy and Okech would wipe everything with wet leaves to make sure it was clean before he ate it. Oh, and there's one other thing! Look at the photos of Okech on the Ngamba Island website. He looks just like the American president, George W. Bush!

Then there were the "sisters," Becky and Sally. They're not sisters really. They aren't even closely related. I remember a big escape once when twenty-three chimps found a weak place in the fence and they all got into the administration area. It was chaos. I arrived at the island to be greeted by twenty-three excited chimps holding all kinds of things—cookies and shoes and clothes and anything else they could steal. With another caregiver, I went walking around the administration area, and everywhere we looked we found chimps. They had raided everywhere, including the drink store. And I could see that some of the beer cans had holes bitten into them. It made me wonder if any of them had drunk some beer. Eventually, we tracked all the chimps down. We found Becky and Sally fast asleep in one of the caregivers' beds. They were snoring and smelled like beer. We carried them back to the holding facility and next morning, oh, the "sisters" had the biggest hangovers!

So, as you can see, I had some wonderful—and sometimes difficult—times caring for the chimps of Ngamba Island. Some of them became my closest friends. You could say they became like members of my family. And family is an important thing, both for human beings and for chimps.

But what does family really mean?

Chapter 8

WHAT FAMILY MEANS

What does the word *family* really mean? And what makes up a family? I know there has been a lot of talk around the world about that question. Many different people have many different opinions, but to me, family is all about one special word . . . love.

Over the years, I've learned that family is more than just those people you are related to by blood. At different times in my life, different people have been family for me. When I was young and living in Burundi, I had my relatives and my friends around me. My mother and father, my six sisters and my two brothers, they were all my family. But things change. Young men grow up. When I married Nowera and we had our two little boys, they became my immediate family. When I was down in Bujumbura, working at the halfway house, it was them I thought of and worried about. And, during the times I was at work or living in the bunkhouse in Bujumbura, Elie was my good friend. Then I left Burundi. I went to Kenya and then to Uganda. Suddenly, I had no country and no real home. I had only one blood relative living with me—my younger brother, Minani. My good friend Elie had gone back to Burundi. I was spending my nights at the drinking club, but those men weren't really my friends. After I stopped drinking, I had my spiritual family at church. And, during this time, Debby Cox became like family. I'm not sure how I would have survived those times without Debby. I've always believed that Debby was my sister.

And, along with all these people, there were the chimps. People sometimes say, "Stany, you helped to save all those chimps." But, you know, the chimps saved me too. There have been times in my life when I have been so lonely. I think back to those days in Kenya, when Debby had gone to Uganda and I was alone—the chimps at Sweetwaters were my only friends. They are so much like humans, and they build bonds of friendship very much like we do. So yes, the chimps have been part of my family as well.

Even though there was so much happening at Ngamba Island and even though I was developing close relationships with people like Debby and my chimpanzee friends, I still desperately missed my wife and my two boys. There would be some dark days, and even darker nights, when I would wonder what I should think about them. I wondered what I should do. Because there was no way I could contact them. There were no telephones, and no way to get a letter to them. I didn't even know where they might be. Were they still in our family village? Had they fled? Perhaps, I thought, the most appropriate thing would be to mourn their loss. I really didn't believe that a young mother and two little boys could survive such as a barbaric war.

By this time, I knew that the Hutu people of Burundi had been massacred in large numbers—tens of thousands. I found it hard to accept that Nowera and my sons could possibly have survived. I wondered if I would ever find out exactly what had happened to them. Was there even the slightest hope that they were alive?

I kept telling myself it would be better to go back. I would try to find Nowera, Charles, and Innocent. And if I found them, then together, as a family, we would accept our fate. Surely, it was better to be with my loved ones and face the danger together. I said to Debby, "I know I will probably be killed, but it is better to be with my family—even to die—than to be away from them and live alone. I want to go back to Burundi."

Debby kept reminding me that I'd been away from Burundi for a long time. It would look to the soldiers or police like I'd been training as a rebel fighter. Why else would a Hutu man leave Burundi for so long and then go back? She'd heard that the Tutsi military were rounding up young Hutu men and imprisoning them. She said I wouldn't even have the chance to start looking for my family. And, everyone thought Burundi would soon calm down because the civil war had now stopped in Rwanda. When Burundi calmed down, Debby told me, it would be safe to go back.

But things didn't calm down. The fighting in Burundi continued.

What I didn't know at the time was that Debby had contacted Aly Wood to see if there was any way she could search for Nowera, Charles, and Innocent. After JGI had left the country, Aly had stayed for a time in Burundi, working for the UN ambassador in Bujumbura. It's hard to imagine how terrifying her life must have been in that city.

"People were being killed all the time," Aly told Debby. "And, even if they weren't shooting at you, anyone could easily be killed by stray gunfire. I remember, when the bullets started flying, I would go to the bathroom and sit in the tub with a radio. In most houses, the bathtub is the strongest and safest place to be. Burundi was just lawless. This was why Stany couldn't go back. Because he had been away for so long, and the war had become so violent, he would have been considered a rebel. He would have been killed. No, it was safer for us to find his wife and kids and try to get them out of Burundi."

In 1999, the search for my family was finally successful. Aly had been in contact with my friend Elie, and Elie was able to find my youngest brother, Nyabenda. Together, Elie and Nyabenda began searching for Nowera and they found her.

My wife and boys had been through a terrible ordeal. After I'd left for Kenya, they were trapped in the middle of a place like hell. Nowera told Elie and Nyabenda how soldiers would come into the village and the people would have to flee. My wife would run with Charles at her side, and little Innocent on her back. She would carry their few possessions in a bag balanced on her head. Along with other families from our village, she would hide in caves down in a nearby valley. The soldiers would come, take whatever they could, and then leave. But sometimes two groups of soldiers would arrive. One group would start to drive the villagers towards a second group, who were waiting to ambush them. The Tutsi soldiers were looking for young Hutu men to arrest or kill, but sometimes they would just kill everyone.

After the initial contact with Elie and Nyabenda, and despite the horrible conditions she was living in, Nowera refused to leave Burundi. Her extended family was in Burundi, and Burundi was all she'd ever known. Aly and Debby made sure they kept in contact with Nowera. They kept gently encouraging her to leave. Then, Debby sent Nowera some money, but that money never arrived. I don't know if someone else from my family took the money, or if someone from Nowera's family took it, or if it was stolen by someone else before it was delivered. Whatever the reason, after this happened my wife decided she would try and leave Burundi.

Following up the information from my brother, Aly went to see Nowera, Charles, and Innocent. They were living in a settlement for displaced people, just outside Bujumbura. Though they were now mostly safe from soldiers, life in that settlement was difficult. Much later, Nowera told me how she'd seen so much sickness and death. There were many people with malaria and cholera and typhoid. Each family would receive a small ration of food—a little over two pounds of rice for a family, which had to last for a week. People in the settlement were like walking skeletons. Every day someone would die. There was always a body that needed to be buried.

At this horrible time, I believe that God intervened for my family. The American missionary I had met when I was a houseboy in Bujumbura was working in that settlement. I thank God for this man. He ran a clinic in the settlement, and through that clinic my wife and my boys were given food and medicine. Because of this, they survived.

Because she worked for the UN, Aly Wood had contacts with government and aid agencies across Bujumbura. She arranged to take Nowera and the boys out of the settlement. With Elie helping to translate her words into Kirundi, Aly completed all the necessary paperwork. It took a long time, and I don't know exactly what the process was, but Aly persuaded the authorities to allow Nowera and the boys to leave Burundi as refugees. They were given a *passage sûr*— this means safe passage—document that allowed them to travel across the border and into Rwanda, where the civil war had ended.

Many years later, Aly retold the story of that journey. She caught a bus with Nowera, Charles, and Innocent. The journey to the Rwandan border took about three hours. Nowera is a short lady, and at that time she was also very thin and frail after living in that settlement. She didn't speak any English. Aly could speak only a little Kirundi, so they didn't say much during the journey. They rode in a small bus, with about twenty or thirty people on board. Innocent had malaria, and he sat on Aly's lap, while Charles sat on Nowera's lap. They had a plastic shopping bag which held all their possessions. Everything my family owned was in that one small plastic bag.

At the Rwandan border, they were stopped and questioned by authorities. The guards wanted to speak with Nowera alone, but Aly told her to stay quiet. Aly told the men at the gate that Nowera and the boys were traveling with her and she was responsible for them. Aly said the border guards lost interest and they allowed my family to cross without any problems. "Okay," Aly remembers saying at that point. "We're out of Burundi! Things will get easier from here."

Nowera and the boys stayed with Aly and her partner, who were now living in Kigali, Rwanda's capital city. After years of violence and fear and disease and discomfort, the luxuries of that home must have seemed amazing. My family ate a meal of beans and rice, they had their own room to sleep in, and the kids were able to have a wash in a full-sized bathtub.

"I'm sure this was the first time Nowera and the kids had experienced these things," Aly said. "You know, a big bed with a thick mattress and real sheets, and a bath in a full-sized tub."

However, my family was still a long way from where I was.

Their next journey was on another bus, this time going from Kigali to Kampala in Uganda. This was a longer journey—ten hours—but Aly wasn't as worried. She knew they were unlikely to be harassed by the authorities in Rwanda and Uganda.

I was working on Ngamba Island the day Debby told me that Aly had found my family. I was very relieved, because I now knew they were alive. I wondered if I would soon be able to travel back to Burundi and see them. Maybe then, one day, I could arrange to come back to Uganda with them and continue my work with the chimps. Then, Debby told me Aly was coming for a visit, and she was bringing something for me from my family. I wondered what it might be. Perhaps they had written me a message or a letter. I felt excited.

Aly, Nowera, Charles, and Innocent arrived in Kampala and then caught a minibus down to Entebbe. Aly remembers the taxi park at Kampala being like a zoo. There were lines and lines and lines of mini buses and it looked like complete chaos. Poor Nowera was nervous. She had never been to a place where there were so many people and cars.

That afternoon, Debby contacted me. She asked me to come back from Ngamba Island and go to Entebbe Zoo the next day. When I asked why, she told me that Aly was arriving in Entebbe tomorrow.

That night I had a dream—I was back with Nowera and my boys.

The next morning was a Sunday, and I went to church. As the service went on, I didn't know what to think or what to pray. What was the day going to bring me? And what was it that Aly had brought from my wife to give me?

After church, I went to the zoo and I waited. It was about three o'clock in the afternoon when I saw some people in the distance. They came walking in through the side gate of the zoo. They started to walk in the direction of Debby's little house, which was on the zoo grounds. I didn't recognize the people.

Then I saw Aly. And Aly was alone. That's when I remembered the dream about seeing my family again. I immediately lost faith in that dream. Aly had come to Entebbe, but she was by herself. Aly came up to say hello to me and then we walked over to Debby's house.

When we arrived, Debby came out of the house and asked me if I would go inside to fetch some chairs. "Let's sit out in the sun and have a catch-up," she said.

So, I went up to the house.

I opened the door.

I was paralyzed.

Nowera, Charles, and Innocent were standing inside.

It had been five years since I had seen my family. Five years, not knowing if your wife and kids were safe, or even if they were alive. Five years apart and now . . .

Oh, how we hugged!

"I'll never forget the look you had on your face," Debby said to me.

Aly started crying.

Innocent was scared. He didn't want me to go near him. He had been so young when I left Burundi, he didn't know who I was. Charles was okay, he did remember me. He was a bit nervous, but not scared. He knew I was his dad.

No, I will never forget that Sunday afternoon. It was a wonderful day. No, it was more than that. It was a big miracle.

After the joy of being back with my wife and boys, it didn't take long for the reality of our situation to hit home. My wages from Ngamba Island allowed me to manage a simple existence on my own, but now I had to provide for my wife and two growing boys. I was renting a single room in Entebbe. All I owned was a small bed, a kerosene lamp, two cups, and two spoons and forks, and that was all. There were no chairs, not even any mats. When I sat, I just used to sit on the dirt floor.

Although Debby didn't own much herself, I can never forget the love she showed to me and my family at that time. She was so kind. So very kind. My family had dinner at Debby's house. After dinner, she said to Nowera and me that we could take anything we wanted from her house. Anything we wanted! Such generosity! We took a kettle and some cooking things, as well as a blanket and a mattress for the boys.

Even with these extra things from Debby, our single room was poor and not suitable for a family. There was no water and only a squat toilet—a hole in the dirt floor. Worse still, the house was near a swamp, and there were many mosquitoes. When you went to the toilet, you had to take a leafy branch in with you to swat them away. My boys were already sick. Innocent still had malaria, and I thought these mosquitoes would only make it worse.

After Debby lent me some money, we were able to rent a small house for three months. I bought some mats to put on the dirt floor. I also bought a bench seat that we could all sit on during our meals. We had one plate. Nowera would cook a meal and serve it onto the middle of the plate and we would all share.

Because Ngamba Island was more than an hour's boat ride from Entebbe, and because I didn't have a boat of my own, I couldn't easily commute to and from work each day. This meant I stayed on the island for several days at a time, and I was once again away from my family. This was hard for me, but it was even harder for Nowera. She was in a strange country where she knew nobody and where she didn't speak the language. I think my wife felt the isolation very much.

Our neighbors would make comments about the "skinny people" living next door. We didn't want to look like we were poor, but it wasn't easy. We tried hard to keep our house neat and tidy, but there was little we could afford. I was making money at work, but it was important to me that I paid back the money I owed Debby. So, most of my income was going back to her. There wasn't much to spare. We started running out of food.

Around the shore of Lake Victoria you would see people selling small dried fish. These are called silver fish. They are caught in the lake and then laid out on large mats to dry in the sun. Mostly these fish are then crumbled up and used to feed animals like pigs or chickens. They were not a very good food for people, but they were all I could afford. To go with the silver fish I would also buy a large sack of cassava flour. This flour was cheap, because it was mostly used to make glue for putting up posters. If there was any money left, I would also buy a bag of charcoal for cooking.

This was a terrible time—knowing all I could afford to feed my family was animal food and paper paste. Still, I didn't tell anyone. In fact, before I shared this story with you, I hadn't told anyone about what was happening. I believed it was my problem.

There was another source of food, of course. Just like at the halfway house, the chimps on Ngamba Island's received a lot of good food. They had millet porridge in the morning and fresh fruit and vegetables throughout the day. Still, even though my own family was eating poorly, I didn't want to take any food from the chimps. It was difficult, but I loved the chimps and they had saved my life. I just couldn't take food from them.

At times like this, it was easy to feel disheartened. I had to remind myself of all my blessings. I had my family with me once again. They were alive. They were safe. We didn't have so much, but we were together. During my nights on Ngamba Island, while the other staff were sleeping or watching TV, I would go up to the feeding platform and look up into the heavens, and I would pray to God. I thanked Him for bringing my family safely back together again.

Sometimes, however, the burden of responsibility felt like too much for me to bear. I remember times when I had no money—I mean not even one coin in my pocket. One Sunday, I spoke with a lady at church. I told her that I had some problems. I thought that maybe she would have advice for me. She told me there were people who would lend money but that they would charge interest, which meant I would have to pay back a lot more money than I borrowed. I remembered reading in the Bible where it said, "Do not lend people money for profit." So, I told myself I wouldn't do it.

I remember walking home from church that Sunday. I was disheartened. As I walked, I said to God, "You have given my family back to me, but now I cannot care for them properly. So, if I can't feed them, it must be your fault."

I was filled with sadness and shame.

I got home and began to pack a small bag with some of my things. "I cannot care for you," I told Nowera. "So I'm leaving." And I did . . . I left. I went out to live on Ngamba Island.

Nowera said nothing in reply.

It took a long time, but my situation slowly improved. And, you know, every day was not the same, but I kept trying. I sent what I could to Nowera. Eventually, I went home again. When I had spare time at Ngamba Island, I used to try and catch fish so I would have better food to bring when I went home. I kept working, and I kept trying. But it was so very hard.

I knew Debby would be willing to help me if she knew how bad things had become, but I tried to keep our relationship purely professional. I didn't want to burden Debby or Aly with my problems. They had been so kind to me. They had already done more than anyone could ask or expect. They had delivered my family back to me. And I knew they had problems of their own to worry about.

For my wife, these early months in Uganda were especially difficult. When she arrived in Entebbe, Nowera had only one set of clothes. The boys too. They would be almost naked at times. Wilhelm, the boss of Entebbe Zoo, and his wife were very kind. They gave me some clothes to give my family. But the clothes didn't fit my wife. It took many months before she would have another set of clothes to wear. This meant she wore the same clothes every day for months, and it filled her with shame. She would spend the day hidden away in our house. She wouldn't even come to church with me.

Our eldest boy, Charles, was a wonderful help during these times. He learned to speak Luganda quickly. So, he would go out for Nowera when she needed something. Nowera would tell him what she needed, and Charles would go out and get it! Charles was only six when he arrived in Uganda, but he proved to be a fast learner. He had to repeat one year of school, because he was behind the other kids his age, but after that he passed every year. Debby paid for both Charles and Innocent to go to a private school. Can you see how very generous Debby was to my family?

And my boys proved to be smart! They both did well in school. They were both able to go to university.

It isn't only tribal and ethnic differences that create hardship for so many people in Africa. Religious divisions lead to despair for many people—in Africa and around the world. I believe that people of faith, no matter what that faith is, have a duty not to hate, but to show true love. We are to bring justice and hope. We must shine like stars in a dark world.

It was the early 2000s. My family and I were moving into another rented house. As we were arranging to rent this new house, the landlord warned us about the lady living in the house across the road. That lady was very sick, and we needed to be careful because no one knew what type of sickness she had. This lady also had a baby. And because the lady was so sick, she didn't seem to be caring for her baby.

After we moved into our new house, Nowera saw the baby, naked and alone and crawling outside the lady's house. So Nowera went over and picked up the little girl. She brought the baby to our house, washed and cleaned her, and gave her some food. After she had cleaned and fed the baby, Nowera went to the lady's house looking for the mother. She told me the house was dirty and the smell was horrible. Nowera found the baby's mother—she was named Joy—and noticed how sick she was. Joy had weeping sores all over her legs, and in some places, the skin was peeling away.

Nowera asked Joy who was there to help her. Joy replied that she had a husband, but that she was a Christian and her husband was a Muslim. After their little girl was born—they named the baby Shadia—Joy's husband declared that the baby was not his and he left the home. Joy became more and more unwell.

Local people were unhappy with Nowera. They told us to stay away from Joy, because her disease was probably contagious and we would spread it around the neighborhood. And, of course, there are many terrible diseases in Africa. Viruses like HIV/AIDS, Ebola, and Zika all began in Africa.

We would not walk away from our neighbor. Nowera began to clean Joy's clothes and her home. I took Joy to some hospitals in Kampala, and, even though the doctors did many tests, they could not find out exactly what was wrong with her.

Eventually, with love and care, Joy returned to good health. However, she could not afford to properly care for Shadia. So, Nowera and I decided to adopt Shadia as our daughter. To this day, she lives with my other children. She is part of my family. And Joy is our sister. She comes to visit my family often.

But, you know, even though Shadia has now grown into a young woman, people still ask hurtful questions. I see people looking at my family, and they look at Shadia and realize she looks different to my other kids, and they make comments. I have heard people say I had Shadia after I slept with another woman—that Joy was my mistress.

They thought I had deceived my wife. Nowera knew the truth. She didn't believe this. But, there was more.

People would ask, "Why did you adopt Shadia and care for her mom?" They would say, "You put your own family at risk." "You acted foolishly."

No one knew what sickness Joy had, so it was true that there was a possibility my family would get sick as well. That we would catch Joy's illness. But for me, I had read stories about Jesus in the Bible. Jesus was with the sick people and He cared for them. Jesus never got sick. If we say that we follow Jesus, then I believe we must do the same as He did. We must show true love.

During this same time, there was a tragedy unfolding in my family. The brother who was born after me was named Gregory Minani— "Minani"—meaning the child born eighth. But, unlike me, nobody used his first name. Nobody called him Gregory. He was always called Minani. While I worked with the chimps at Entebbe Zoo and Ngamba Island, Minani worked as a housekeeper for Debby Cox, as I told you earlier.

Debby says, "Minani was such a wonderful guy. He was honest and hardworking. He was reliable too. If you asked him to do something, you knew it would be done. And he always did what he said he would do. Both Stany and Minani were such outstanding young men."

But, as time went on, Minani's life began to take a bad turn. During the early days in Uganda, the two of us were drinking heavily. But, at the point when I found the church and walked away from my damaging life, Minani continued on his dangerous path. He met a woman and began to go out with her, and she took him to unhelpful places. She wasn't good for him. It wasn't just alcohol now. They were both using drugs and doing other dangerous things. And, you know, my brother and I didn't know much about HIV. When we were in Burundi, we didn't see much evidence of the disease. But, in Uganda it was different. There was more freedom and HIV was a big problem.

Debby also knew what was happening with Minani. She saw his life begin to slide. She tried to speak with him. She even gave him condoms to protect himself, but he wouldn't listen.

Minani married this woman. They had kids. Their first two children were girls, named Susan and Denise. They were happy and healthy kids. But at some point Minani and his wife contracted HIV, and their third child, a boy they named Steven, was born HIV positive. Although I was struggling to pay for the needs of my own family, and we were also caring for Shadia, Nowera and I knew we had to take care of my brother's three kids as well. My brother and sister-in-law were still alive at this time, but they were very sick and couldn't care for their kids properly. Nowera and I took Susan, Denise, and Steven into our house.

Once again, people were unhappy with our choice. They said to us, "You are being foolish. You're putting your own kids' lives at risk! What if they catch HIV?"

"No," I said. "Jesus cared for the sick, and that's what we are going to do as well."

Minani and his wife eventually developed AIDS, which is the deadliest form of the HIV virus. They both died. And I remember the moment my younger brother died. Gregory Minani passed away in my arms.

As well as adopting Shadia and Minani's three children, Nowera and I had two more kids of our own. We had a boy we named Gideon and a girl we named Christine. Our family, now numbering eight, would grow up together and live as one. We were all one family. Yes, different parents, but integrated as one. So you see, I know how to do integration—with chimps and with people!

After the war finally calmed down in Burundi, I went back for a short visit. This was around 2005 and I wanted to find out what happened to the rest of my family. I wanted to see if any of them were still alive. That short visit to the heart-shaped land broke my heart.

I found out that all the people in my village had been killed. At different times, the Tutsi soldiers came into the village to plunder and steal and to take away the young men. On one occasion, my father had been working out in the family garden. Even though he was not a young man, and even though he'd been a policeman working for the Tutsi authorities, they shot him dead.

The soldiers went to my parents' hut. My mom was inside the hut. She was sick and couldn't run away. The soldiers tied her up and then set fire to the hut. My little family home was burned to the ground. My mom was still inside.

There was more. I spoke to some people, and they told me what happened to my youngest brother, Nyabenda. He had been caught by soldiers, who found he was carrying money. The soldiers stole the money and then they tortured him. And then, my youngest brother—who, with Elie, had helped Aly Wood to find Nowera and my two sons—was tied to the back of a truck and the soldiers drove around the village with him attached. He was still alive when they began to drive. And they dragged my little brother around behind that truck until he was torn and bloodied and had died.

Today, my family and I rarely speak about the war. Most of our kids are Ugandan, and the war is not relevant to their lives. My two oldest boys, though, were born in Burundi. But still, we don't discuss it. Innocent was too little to remember, and I don't speak about it with Charles either. Even though he was old enough to remember some things, we don't talk about them. I don't want his mind to be poisoned by those memories. I don't want him to live as a man holding onto hate.

Not every young Hutu man was killed in Burundi. My dear friend Elie Nkurikiye survived the war. I will always remember when Elie traveled to Uganda to see us. He visited Ngamba Island with an American man named Jim Kurtz, who was a longtime supporter of JGI and became a friend of mine. It was Jim who was with me on

Ngamba Island that time Eddie escaped—when I made him a cup of tea.

After he arrived on the island, Elie went over to the electric fence to spend some time watching the chimps in the forest. To most of the chimps, Elie was just another visitor, but there were two chimps who knew him. These two were named Umutama and Umugenzi. They were among the chimps we evacuated from the halfway house in Burundi, and after living at Sweetwaters in Kenya, they had been moved to Ngamba Island. Though they had not seen Elie for many years, as soon as they saw him they ran to the fence to greet him. It was beautiful to watch. Elie was always such a wonderful friend. It was Elie who helped to find my wife and boys in Burundi. I love him like a brother.

And chimps make wonderful friends as well. They can be your friends for life. I went back to see my old friend from Burundi, Max. He was still living at Sweetwaters sanctuary, though he was a big male by then. And even though I hadn't seen him since he was little, Max still remembered me. He started pant-hooting as soon as he saw me. I could tell he was glad that his friend had come back!

Of course, among all these friend and family members—both chimpanzee and human—there is one who holds the dearest place in my heart.

My Nowera.

Not only was she able to survive the horror of the civil war and genocide on her own, she has also been able to care for our large family during the long periods when my job takes me away. My wife, she is a tough and strong person. We trust each other. Our love is strong. She understands what I do, and she supports me. What a blessing my wife has been, because sometimes, my work has taken me far away from her.

Sometimes, I have traveled a whole world away.

Chapter 9

CULTURE SHOCKS, CANDY, AND COMMENDATIONS

I've been a very fortunate man. I have traveled to many different countries and visited some of the world's most famous cities. And, just as my first visit to big city Nairobi opened my eyes, those journeys beyond Africa's shores were going to be a big adventure. Traveling overseas would introduce me to new sights and experiences and many kinds of Western "magic."

Australia was the first country I visited outside of Africa. I suppose this isn't surprising, considering my close relationship with Debby Cox, who is an Aussie. I had also worked with many Australian volunteers, both at the halfway house in Burundi and on Ngamba Island. And there has long been a connection between Taronga Zoo, which is in Sydney, and the work of the Jane Goodall Institute.

Australia was really interesting for me. It was my first chance to see how people lived compared to Africa. This was in the year 2000. I did a keeper exchange with someone from Taronga Zoo. One of their keepers went to Ngamba Island for three months while I was in Australia. In addition to the zoo in Sydney, I went out to visit Taronga's open range zoo, which is in a place called Dubbo, about 250 miles west of Sydney.

As well as Sydney and Dubbo, I had the chance to visit other parts of Australia. I went touring around with Debby and her mother. Debby was visiting other zoos to see how they cared for their chimps. I remember we visited a zoo in Queensland—it was in a large botanical garden—where they had a small group of chimps. During this trip I had my first chance to see many different animals. I remember the first time I saw kangaroos. And, can you believe, this was also the first time I stood by the ocean. I knew all about Africa's freshwater "oceans"— Lake Tanganyika and Lake Victoria—but that time it was the Pacific Ocean. I remember smelling the salty sea air. And I saw dolphins!

However, before any of this happened, my arrival in Australia was very confusing. In the year 2000 Sydney hosted the Olympic Games, so Sydney airport was a busy place. When I came out of the plane, I couldn't believe how many people were in the airport building. It was like walking through a maze. It was so easy to feel lost.

I was supposed to be traveling from the airport to Taronga Zoo, which was about fifteen miles away. But, there had been a mix-up about what time I would arrive. The people at Taronga thought I was arriving on a different day. So, after I went through customs, I walked out of the terminal building, and all I could see were people and buses and taxis and long lines of cars. But I couldn't find anyone who was waiting for me. I only had a little money in my pocket, which I didn't think would be enough for a taxi. And, even if I started to walk, I didn't know the direction to Taronga Zoo. So, I found a seat outside the terminal, and I sat down and wondered what I should do.

After I had been sitting for a while, I remembered I had a business card in my wallet, which had been given to me on Ngamba Island by a lady who worked at Taronga Zoo. I found a pay telephone, and someone helped me to ring the lady. I told the lady where I was.

"It's okay," she said.

I wasn't sure what she meant. Was she coming to collect me? Would she tell someone else? I didn't know what was going to happen. So, I went back to the seat outside the terminal building, and I sat down again.

Then, after a short time, an Australian policeman walked up to me. "Oh, hello there," he said with a smile. "You must be Stany."

I couldn't believe it. How could he possibly know my name? I thought that Australian police officers must have some sort of magic!

Actually, the lady I'd phoned had then rung Taronga Zoo. Someone at Taronga had rung the airport police and asked if an officer could come and find me to let me know that a car from the zoo was on its way. They didn't want me to worry and get myself lost. Eventually, a car from the zoo arrived, and they gave me a lift.

Taronga Zoo has cared for chimpanzees for almost a hundred years. Today, the zoo has a large community of more than twenty chimps. They are all different ages, from sixty years old to infants. The chimps at Taronga live in an outside enclosure surrounded by high brick walls and a deep moat. The new Entebbe Zoo chimp enclosure had been modeled on the one at Taronga. From their enclosure, the chimps at Taronga have a beautiful view over the famous Sydney Harbor towards the Opera House and the city. Some of the keepers told me they had chimps who liked to sit out in their exhibit on New Years' Eve and watch the fireworks display over the harbor. I thought Taronga's chimps were so lucky.

At Taronga, the chimps were largely self-managed. This meant the rank and status of each individual was determined by the chimps themselves. The keepers cared for the chimps in a way that respected the group hierarchy. While I was working with these chimps, I had the chance to learn new methods of chimp management and new enrichment ideas. I also had the chance to share some of my experience and skills.

Louise Grossfeldt was working as a Taronga Zoo chimp keeper back in 2000, and she has worked with chimps for about thirty years.

Louise also worked with me and Debby in Uganda when she came to Africa for a visit. Today, she is a curator at another zoo in Sydney. Louise told a friend of mine, "I've never met anyone like Stany. The way he communicated with the chimps . . . I mean, I'm convinced he can speak 'chimp.'"

I learned much at Taronga Zoo. I worked with the staff of the African division. These people looked after the chimps as well as other African animals like giraffe and zebra. At Taronga I learned a lot about following rules. In Africa, we didn't have so many work rules, it was very open, but in Australia . . . oh, there were so many rules to follow! Sometimes, I would get in trouble for doing things that weren't in the rules. But, I also learned new ways to work with chimps when there is not so much contact with the keepers. At Taronga Zoo, there was always some kind of barrier between the chimps and the keepers, unless the chimps were sedated. This meant the chimps worked out a lot of things for themselves, and the keepers had to work with the chimps to get their cooperation. So, the way the keepers trained each other, and the way they trained the chimps, was quite different from what I was used to. This was very interesting for me to watch and learn.

While working at Taronga Zoo, I had the chance to learn more about Australia. Some of the things I learned were surprising. They seemed more like magic to me.

One time, I was walking down the street with a friend in Sydney. The city was full of people and traffic. It was a hot day. We walked past some large machines. One of these machines was bright red and had "Coca-Cola" written on it. The other had a glass front, and you could see chocolate and candy inside. I had never seen these machines before. My friend tried to explain how they worked, but I didn't really understand. So, she put some coins into the red machine, and there was this whirring sound inside and then, *bang!* And my friend reached down and pulled out a can of cold soda from the bottom of the machine. Then, she put some coins into the candy machine, and it made another whirring sound and then a chocolate bar fell to the bottom, where she could reach in and pick it up. I couldn't believe it. I mean, if you paid these machines some money, they would give you soda and candy. The machines were shops! And I wondered, was this some type of magic?

It wasn't only the shopping machines that surprised me. I also had an opportunity to experience the Aussie approach to work. Time management in Australia seemed very strange to me. Australian workers would be working at their jobs and then, when it came to be a certain time, all the staff would stop their work to have a break. They did this even though the jobs they were doing weren't always finished. They would go to the kitchen or to the shops and get some tea or coffee and maybe a cookie or some cake, and then, when it came to be another certain time, they would go back to doing their jobs again. In Africa, we would always finish our jobs first, and then we would have a break. In Africa, it was not so much about what time of the day it was, but what you were doing at that time. When I was in Sydney, I learned how important the time of day was to Australian workers.

While I was at Taronga Zoo, I had the chance to attend an international conference on animal enrichment. Enrichment is an important part of caring for animals in zoos and sanctuaries. And it's really important when you're caring for intelligent animals like chimps. At the conference, I learned how enrichment tools and techniques stimulate natural behaviors and give animals an opportunity to live in a way that is more like their true nature. So, at Taronga Zoo they had built a concrete termite mound in the chimp enclosure. This mound had holes in the side and these holes connected to containers filled with things the chimps loved—like fruit jam. The chimps learned to poke a stick into the holes, and then they could lick the jam off the end of that stick. This allowed them to display tool use, just like wild chimps do when they "fish" for termites using sticks or stems of grass.

There were people visiting this enrichment conference from around the world. I was able to hear about new techniques I had never seen before—not just about giving food, but ways to stimulate the animals' minds. It was very interesting. Some of these enrichment tools were good to use in my work, such as new items to use in my "jackpot" technique. But some were not relevant to chimps living in a forest sanctuary like Ngamba Island. Anyway, I enjoyed this conference. And I enjoyed my time in Australia very much!

Europe was a much different place than Australia. I first went to Europe in 2010, and once again this visit was part of a keeper exchange program. The first exchange was between Ngamba Island and Chester Zoo in the United Kingdom. On this exchange, I traveled with another Ngamba Island caregiver named Amos. We stayed in the UK for two months.

In some ways, my time in the UK proved to be more difficult than my visit to Australia. Chester Zoo is a very old zoo, and I think they must have had processes in place for a very long time. I didn't feel like I was much use there. And I didn't feel it was my place to say very much. The Chester keepers were very good, and they were nice people, but I think they were set in their ways and really didn't need my advice.

Like Taronga, Chester Zoo has a big chimpanzee community. And like Taronga, their chimps are housed in an outdoor exhibit, with a heated indoor enclosure that allows them protection from the cold English winter. The Chester Zoo chimps were important to the future of wild chimpanzees. There are four subspecies of chimpanzee, and most zoo chimps carry genetics that are a mixture of these subspecies. However, a study showed that some of the Chester Zoo chimps carried only the genetics of the critically endangered western chimpanzee. This was such an important discovery. To help save this subspecies, Chester Zoo started a breeding program for their western chimpanzee.

In addition to Chester Zoo, I have memories of some other things in the UK that were rather different from anything I'd ever seen in Africa. I remember visiting Blackpool, which is a town on the coast of the Irish Sea. Many English people go to Blackpool for their holidays. When I was there, I saw a horse walking along the road. Now, there was a man riding on this horse, but when the traffic lights turned red, the horse seemed to know that he was supposed to stop! The man didn't even need to speak to the horse. The horse just stopped by itself. And I wondered, how did that horse know to stop at the red light? I think he must have been a very smart horse.

Also, while I was at Blackpool, I saw a wedding down on the beach. It was the weekend, and there were a lot of people down at the ocean. As we went closer to the beach, I realized all the people at the wedding were men. Some of the men were wearing skirts. These skirts were not like Scottish kilts but the kind ladies wear. And then I noticed that the bride and groom were both men. I was confused. I asked my friend what was happening, and he told me it was a gay wedding. Now, in countries like Uganda, homosexual acts are illegal, and a gay wedding is not something you would ever see.

After I was in the UK, I traveled to Germany and spent two weeks at the Max Planck Institute in Leipzig. Researchers at the institute were studying chimp behavior, and some of the team had come to work at Ngamba Island, so we knew them and they knew us.

Doctor Juliane Kaminski was a research group leader at Max Planck. Part of her work was to conduct research on the ways chimpanzees use their brains to think and learn and remember. Besides the work she was doing in Germany, Doctor Juliane had come to Ngamba Island, where she worked with me and some of the chimps.

Juliane said, "Stany's help was crucial for us when we were working with the chimpanzees. I remember how we would give the chimps problems to solve, and their solutions to those problems would tell us something about how they think, and how they see and understand the world around them. Sometimes the tasks we set up required the chimps to be in a specific place, and Stany would help us make sure they were where we needed them to be. It was wonderful to watch him work and to see how the chimpanzees completely trusted him. Stany was also a great help when we were setting up our work. Sometimes there might be things that we overlooked, and he would point out that our ideas wouldn't work and then help us find a better way of setting things up so it was easier for the chimpanzees to understand. Stany could understand the chimps in a way that others couldn't. Plus, he was always so cheerful and fun to work with."

I liked working alongside Doctor Juliane. She was such a nice person, and I learned so many things from her and the other Max Planck people, such as the way the mind of a chimpanzee works. Besides visiting places like Ngamba Island, Doctor Juliane and the other researchers worked at the Leipzig Zoo, where a community of chimps lived. This zoo also had bonobos—which are sometimes called pygmy chimpanzees—as well as gorillas and orangutans.

I loved the Leipzig Zoo and the chance to study those other apes. I would walk around the exhibits and watch the chimps, bonobos, gorillas, and orangutans going about their business. And even though I wasn't in Germany to share my experience—I was there just to observe—it was very interesting and I learned many new things.

The city of Leipzig was also interesting to me. We were told it had been the second-largest city in Communist East Germany. The city had some beautiful old buildings made of stone. Some of them were painted in bright colors. But then there were the plain concrete tower blocks, which were built during the Communist years. They were such a big contrast, unlike anything I'd seen before. We were given a guide to take us around Leipzig, and that was fascinating. In their city, they maintained the history very well. I remember there was a Catholic church that had been turned into a museum, and they were charging people to go inside. I couldn't believe it. I'd never seen that before either—a church that made people pay money just to go inside.

In addition to the history, Germany was a very modern and efficient country. They had these fast trains, which could take you across the country, or to other places in Europe. The train stations were as busy as airports, and those trains look like space ships. They went so fast—over 180 miles per hour!

The language barrier in Germany was a challenge for me. I speak five languages, but I couldn't speak any German. At first I thought the people didn't like me. I would say something to them, and they would walk past without saying much in reply. Many German people can speak English, but my English was not so good, so I don't think they could understand me. When we were at the zoo and Max Planck Institute, it was okay, because almost everyone could speak very good English, and we understood each other.

The United States is a place I have visited a few times. I have traveled to America to share my stories, to visit other chimp sanctuaries, and as part of keeper exchanges.

Jim Kurtz remembered one of my first arrivals in America. "I remember seeing you in the crowd," he told me. "You were traveling with a suitcase, but it didn't really have anything in it, because you carried all your stuff in a plastic grocery bag."

I remember staying at Jim's beautiful home. He had a refrigerator in his house that could do magic. His refrigerator was like those shopping machines I had seen in Sydney. There was a square hole in the front of Jim's refrigerator, and when you put your glass into that hole, the refrigerator would give you some ice for your drink. You just put your glass in, and out would come the ice cubes. Magic!

One of my other memories of America was the huge department stores. Jim took me to a big store not far from his house. Oh, that store was unlike anything I'd ever seen—in Europe or Australia and certainly not in Africa. That store looked as big as an entire town. Inside there were shelves that stretched away as far as you could see. And they had everything on those shelves—food and toys and clothes and shoes and anything you could imagine. You could buy anything at that store.

"Careful," Jim said to me while I was standing between all those shelves. "You might catch flies in that mouth." My mouth was hanging open because I was in shock.

On my first visit to America in 2006, I traveled to Disney's Animal Kingdom in Florida. Disney had a relationship with Ngamba Island through the Jane Goodall Institute. Though Animal Kingdom didn't have chimps, they did have a group of western lowland gorillas. Disney was such a busy place. I couldn't believe how many people were there to visit the animals. So many people in one place!

I went to Disney to learn about new management processes. One of the things I wanted to learn was how the caregivers were able to ask the gorillas to hold up their arms for injections or to take blood. They didn't have to sedate the gorillas or create a diversion. They gave a command and the gorilla did what they asked. And, as a reward for doing the right thing, the gorilla was given a treat. I wondered if this would work for our chimps at Ngamba Island. When I got back to Uganda, I practiced what I had learned at Disney. I did this by working in the holding facility with chimps like Mawa.

Besides Disney, I visited two important chimpanzee sanctuaries in the US. Chimp Haven was in Keithville, Louisiana. Founded by a lady named Linda Brent, Chimp Haven is the largest chimpanzee sanctuary in the world. There were hundreds of chimps living there when I visited, and I think there are even more today.

Most of the chimps came from laboratories, where they had been used for medical research, including tests and treatments for HIV, diabetes, and other human diseases. When they went to Chimp Haven, each new chimp was integrated into a social group, and they were allowed to build relationships at their own speed.

The established chimp groups were allowed into outside enclosures, and they had access to more than two hundred acres of forest and woodland. This was good for them. After so many years of cruel treatment, I think those chimps were given a taste of paradise.

I enjoyed my time at Chimp Haven, but it was sad as well. Many of those chimps were different than other chimps I had seen or worked with. Many of them had big mental problems, and some really hated humans. This wasn't very surprising, because I knew people had done so many bad things to them for so many years.

I had the chance to meet and communicate with some of the residents at Chimp Haven. This was very special. I remember one time—I can't remember the chimp's name—but I remember he was huge. He weighed about 260 pounds. A few minutes after we were with that chimp, there was a call on the radio to say he'd died. Just like that—dead. Afterwards, they did some tests and found that the chimp had a heart attack. This was the first time I had seen a chimp die so suddenly. I suppose it was just the same as when humans have a bad heart attack, you know: it can happen very quickly.

I was also fascinated by the staff at Chimp Haven. The people working there were almost all women. One time I asked why there were not so many men, and they told me they found it hard to get men to come and do this work. I don't really know why that was the case. In all the other zoos and sanctuaries I had visited, both men and women did those jobs.

Anyway, I really liked Chimp Haven. I was there for a month. The staff were so friendly and kind to me. I think we all learned many things from each other. Linda Brent even asked me to give a talk. This was a very big honor for me, because she was the big boss and she had started Chimp Haven in the beginning. I felt sad when I heard the stories about the chimps and realized how much they had suffered, but I was also happy that Linda and her team had given them a chance to live out the rest of their lives in such a good place.

Another US sanctuary I visited was run by the Center for Great Apes. It was in a place called Wauchula in Florida. Founded by a lady named Patti Ragan, the sanctuary cared for chimps and orangutans.

Many of these individuals had once been "pets" or had been in circuses or movies or entertainment. There were more than forty residents living in beautiful enclosures surrounded by tropical gardens. The sanctuary had also built mesh-covered raceways that went out and through the gardens. This let the chimps and orangutans leave their enclosures and explore the gardens in a way that was safe for both them and the humans.

One of the most famous residents of the Center for Great Apes was Bubbles, who was the first "pet" chimpanzee owned by the singer Michael Jackson. I remember hearing a story about Debby Cox meeting Michael Jackson when he went to see the chimps at Taronga Zoo. Debby was angry at Michael, and she told him he shouldn't keep chimps as pets. At that time, Michael Jackson was the biggest star in the world, and he used to have bodyguards with him all the time, but that didn't stop Debby. Debby is fearless. Debby is like a lioness! When I met Bubbles at the Center for Great Apes, he had grown to be a very impressive male chimp. He weighed over 180 pounds. He was aggressive, powerful, and smart. Bubbles had become the alpha male of his group at the sanctuary.

I also remember another chimp at the Center for Great Apes. Her name was Toddy. She was very old. She had been stolen from her dead mother many years earlier in the Congo. She went to the US and was a performing chimp until she was rescued and taken to the center. I'd never met Toddy before. I crouched down beside the green painted mesh of her enclosure, and I began to softly call to her. I made grooming calls. Toddy came over to where I was, and she started making grooming calls back to me. Then, we sat down together and groomed each other through the mesh. We were strangers, but I could tell that she wanted to make friends. Toddy was a lovely girl. I heard later that dear old Toddy died in 2017. She had a heart attack.

There have been some Africans who have become famous across the world—you know, people like Nelson Mandela, Reverend Desmond

Tutu, and Kofi Annan. There have been some Africans sporting people that the world knows—runners like Haile Gebrselassie and Frankie Fredericks. But in African conservation, it's been non-African people like Joy and George Adamson, Dian Fossey, and Jane Goodall who have been world famous. This was one reason I was so honored to win some international awards for my work with chimps. It was a great blessing as an African man to realize that other conservation people knew what I did and had followed my work.

In 2008, the Disney Conservation Fund presented me with a Conservation Hero Award. These awards recognize individuals and groups who work to save animals and work with local communities to encourage conservation. The 2008 award was for my work to save chimps in both Burundi and Uganda and to acknowledge my skills as an expert on chimp behavior. Disney is such a big organization. To receive an award from them was such an honor.

The Pan African Sanctuary Alliance (PASA) is an association of more than twenty sanctuaries across thirteen African countries. These sanctuaries work to save primates and the wild places where they live. In 2011, along with Ngamba Island, I was awarded a PASA Award in recognition for our outstanding work with chimps. The award was also an acknowledgment of the sacrifices that my family had made on behalf of chimpanzees.

That same year, I won the Carole Noon Award for Sanctuary Excellence, which was awarded by the Global Federation of Animal Sanctuaries. The award I received was to acknowledge my chimp integration expertise, which I had been able to do during my work on Ngamba Island.

One of the most inspiring and humbling places I have visited is the United Nations headquarters in New York City. Doctor Jane Goodall is a United Nations Messenger of Peace, along with movie stars like Michael Douglas and Leonardo DiCaprio. The UN Messengers of Peace are famous people who have agreed to work with the UN and help to focus people on efforts to make our world a more peaceful place.

Jim Kurtz and I were invited to a UN event as guests of Doctor Jane. She gave a speech, and Jim and I were able to go and listen to her. I can tell you, that was such a big honor!

The conference was held in a meeting room at the UN, and a number of famous people gave speeches. During her talk, Doctor Jane gave the room one of her chimpanzee pant-hoot calls, *whoooo-haah . . . whoooo-haah . . . whoooo-haah . . . whooo whooo whooo*. Doctor Jane can do a wonderful pant-hoot! Unknown to those people in the room, Doctor Jane had arranged for me and Jim to reply. So, as soon as she finished her call, Jim and I jumped to our feet and we started pant-hooting back to her. Can you imagine, there we were in the middle of the United Nations building in New York City, two short men—one from America and one from Burundi—and we were pant-hooting like chimpanzees!

"Oh man," Jim said to me after we finished, "that was really something."

After the speeches—and the pant-hooting—we were invited to a ceremony out in the gardens of the UN compound. We were guests of the UN secretary-general, Ban Ki Moon, and in the company of other distinguished guests. Jim kept pointing out famous people to me:

"Hey, that's Michael Douglas over there."

"And hey, there's Forest Whitaker."

And there I was too—a man who had been born into grass-hut poverty in Burundi and survived one of the bloodiest civil wars in history—standing in the middle of all these rich and famous people at the United Nations. It felt like my life had reached a mighty summit. It felt like what I was doing really mattered.

Perhaps it was just as well that, during those happy times, I couldn't see what was coming around the corner. There was soon to be another time of despair and loss. And once again, my family would face much fear and uncertainty. Yet again, my will, my honor, and my faith were about to be seriously tested.

Chapter 10

WHAT IS JUSTICE?

Some stories are complex, and this is one of those stories. Even today, there are parts of this story that I don't know everything about. And even today, I have questions—about motives and intentions, and about fairness and justice.

So, I will try and tell this story as I remember it.

It was 2007, and I was climbing towards new personal and professional heights. My work, and the work of the Ngamba Island Chimp Sanctuary, was being recognized across the globe. People had come to the island to make documentaries and films. Experts were coming to see the chimps and conduct research on them. Debby Cox had created something very special on that small island in Lake Victoria. It was a sanctuary that set a benchmark for others to follow, and I was very honored to be part of it.

This was also a time of change in my personal life. Nowera and I had moved our family from Entebbe to Kawuku, a lakeside village on the very outskirts of Kampala. Kawuku was typical of many Ugandan villages, with its dusty bitumen and red dirt roads, fields of fruit trees, and small clusters of houses, with kids and dogs and wandering farm animals. Nowera and I had purchased a plot of land in Kawuku, and we wanted to live closer to where our land was located, but, although we owned the land, we'd not built a house on it at that stage. So, we had to rent another house when my family first moved to Kawuku, until our new house was built.

As part of our relocation to the new village, Nowera and I decided also to change churches. One of the pastors from the Entebbe church had started a new congregation in Kawuku. He told me of his dreams for a new faith community in the area. He wanted my family to be part of that growth. I was enthusiastic about his plans. I was made an elder at the new Kawuku Miracle Center Church.

My kids, however, were not so happy with the move. The Entebbe church had a very strong children and youth program. My kids had many friends at that church, but, because the church at Kawuku was new, there were not so many kids coming along, and my kids were a bit lonely. They missed their friends.

The early months are often difficult for a new congregation, and it was no different for our new church at Kawuku. We had few regular church members, which meant there was not so much money to cover costs. This meant our pastor was earning little income, and he had a family to support. Nowera and I were happy to share some of our food with the pastor and his family. And while I was barely earning enough to cover our expenses, I also loaned a little money to the pastor to pay for school fees and shoes for his children. I didn't worry too much about this money. The chance to share some of my blessings made me feel happy. I felt this was God's heart.

As part of church planning, the pastor began to speak with me and the other elders about developing new programs. He wanted to create a facility that would assist local children without families. As with so many African communities, Kawuku was home to large numbers of orphans. Uganda had thousands of refugees, from wars in Burundi, Rwanda, and Congo. Some of these refugees were kids who had lost their parents. Also, the impact of HIV/AIDS had left many young people in the village without one or both parents. Unschooled and unloved, many of these children were living a tough life. We would see them out on the streets in rags and dust, just trying to survive from one day to the next. To me, they looked like God's children being ignored.

So, I very much felt the calling of God to help as many of these kids as we could. We could give them a chance to experience true love. We could help them to make a new start, get an education, and have some kind of future.

The pastor told me that he was determined to make the church orphan project successful and viable. I truly believed what the pastor told me. I believed that his intentions were honorable. I trusted that God was in the plans.

Around this same time, while I was working on Ngamba Island, I was speaking with one of the sanctuary's many overseas visitors. This man spoke with me of his passion for Africa, for the beauty of the landscape, and how he wanted very much to help the people. So, I shared with him the story of the Kawuku church and our plans for the new orphan center. I told him how I believed that education for young people in Uganda was essential if the country was to have a prosperous future. I also said the future of conservation relied on local young people being better educated and informed. Over time, communication between me and this man continued. He made some commitments to me, and I gave some commitments to him. We made further plans. And this man donated numerous sums of money to the Kawuku Orphan Project.

But then . . . nothing happened.

After a while, I started to realize that the donated money was being spent, but I couldn't see any real progress being made with the orphan project.

People began asking questions. Where had the donated money been spent? And for what purpose?

I remember the man who made the donations also contacting me. He started asking questions about the orphan project. He wanted to know whether land for the center had been bought. I tried to contact our pastor, but I was told he had gone somewhere else, a place more than six hours away from Kawuku. I was told he had gone to this other place to buy another piece of land, and that the pastor had also used some of the money to buy other things and to start his own personal loan business.

Eventually, along with the other elders, I had the chance to confront the pastor. It was an unpleasant time because he became very angry with us. He denied that any of the money had been used for other things. But, you know, the money was gone, and there was no orphan project starting in Kawuku. So, where had the money gone? I was really upset about all of this, because the money had been spent, but the kids in Kawuku were not getting what they had been promised. I was disappointed and confused.

I felt that the truth needed to be told—to everyone. So, I contacted the person who donated the money, and I told him the truth. I told him that the money was gone.

And then . . .

I don't know exactly what happened, and who did or said what—for many people seemed to have been involved—but there appeared to have been a conspiracy against me and the other elders. We heard that we were going to be charged for stealing the donated money. I was told that papers were being put together. I think that maybe, there were some people who wanted to ruin my life. Someone said the papers were going to accuse me of stealing the money that was donated for the orphan project and keeping this money for myself. I knew, of course, that this was not true.

Time passed. Things seemed to be getting worse.

One day, someone came and told me I was about to be arrested. "You should run away," they said.

But how could I? My family was in Kawuku. I had my work with the chimps at Ngamba Island. I had responsibilities. And, if I ran, I would have to run forever. Also, I knew that only guilty people ran, and I had done nothing wrong. I had not taken the money. My conscience was clear. No matter what happened, I thought that surely, it is better to die as an innocent man than to run away.

It was an otherwise uneventful day in 2012 when, around midday, a group of police officers appeared at my house in Kawuku. The police surrounded my house. They knocked on the door, and when I opened the door, I was confronted and immediately arrested. The police put handcuffs onto me and then they marched me out to their car, in front of my neighbors and friends. I felt humiliated. I gave my cell phone to a neighbor and told her to contact my wife and tell her what had happened, because Nowera wasn't home at the time.

I was driven to the Entebbe police station, which is a multilevel red-tiled building in a compound a short drive north from the center of the city. Once we arrived at the station, the police marched me inside, I went up to a desk, and an officer told me I was being formally charged. I was asked to make a statement. The police said that I had to give them the names of all the other signatories to the Kawuku church accounts. I did what they told me. This led to those other people also being arrested.

I was put in a cell at the police station. It was tough. I was a bush boy. I was a man of the outdoors. And now I was locked inside a small concrete and steel cage. It was just like what so many of my chimp friends had been through. It helped me to understand their anxiety and fear.

I spent three days and three nights in the Entebbe police station jail. It was a small cell, and everything was made of cement and steel. It was damp and smelly, and there were eleven other people in that jail cell with me. I didn't know what they had done. Maybe they were murderers or robbers. Oh, I was so very afraid.

Eventually, Nowera came to see me at the police station, and the officers told her I was being charged and that she would have to raise four
hundred thousand schillings (around $110 US) for my bail. After this
money was given to the police, I was let out. The police then told me that
the charges had been brought forward by the person who had donated
the money but that he was being represented in Uganda by another man.
I knew the wife of this other man. I knew her from the church.

I was charged with "taking donated money under false pretenses."
The police told me I could settle the matter out of court, or I could
go through the court process. I didn't want to deal with any of these
people ever again. I didn't think that I could trust them. I thought it
would be better to go through the court and let them decide.

Even though I was facing court, and maybe jail, I continued to
do my work with the chimps on Ngamba Island. The peace of the
lakeside, the sounds of the birds and insects, and the company of my
chimp friends were all sources of comfort to me during this time.
They helped me to forget the bad things that were happening.

I spoke to the sanctuary manager, and I was initially given leave
to attend court. The manager told me that a member of the Ngamba
Island team would support me through the process. That was a very
big relief to me.

However, the support wouldn't last. No one could have predicted
how long the court process would run. My guilt or innocence was
going to take many years to finalize.

When the court trial finally commenced, I was formally charged
under S/C 305 of the Ugandan Penal Code Act, which was: "obtaining money by false pretenses." The charge specifically related to an
amount of fourteen million Ugandan Schillings (around $3,850 US),
which was among numerous amounts of money donated to the
Kawuku church for the orphan project. My accusers were asked by
the court to stand up, give evidence, and produce documents to support that evidence.

As the trial continued, it seemed to me that the papers the people produced and the evidence that they spoke in court did not agree. But, even though things didn't seem to be going well for the people who had charged me, I was still very worried. I kept thinking about what it would be like to be locked in a jail cell for months or maybe even years.

Although my future was unclear, I tried to stay calm. I knew I was innocent. You know, none of the donated money had ever been put into my bank account. There was no evidence to show I had received even one schilling. I knew that others were trying to discredit me, not just in court, but also in the church, but I knew I would be okay. I had done nothing wrong. And I believed that God was with me.

A whole year passed.

Although no conviction had been made against me, neither had I been found to be innocent. The case remained unresolved. I was called to court again and again, yet no final ruling was ever made. Concerned by the delay, I spoke with my lawyer, who tried to reassure me. He told me that he would appeal if I was found guilty, but then he also told me that being found guilty might mean going to jail, and maybe I would have to stay there for many years, until my appeal was heard. He told me that the courts had the power to do this.

Eventually, after years of waiting as well as numerous puzzling delays, a verdict was finally delivered. I don't really know what went on during this time—why there were so many delays. The magistrate found there was evidence to show that a sum of money had been donated, and that this money had been misappropriated. However, she also ruled that misappropriation of money was very different from the charge of "obtaining money by false pretenses." In July 2015, more than three years after the matter first began, the charges against me and my three co-accused were dropped. I had no case to answer . . . at least as far as the Ugandan justice system was concerned.

Beyond the courtroom, however, the matter was far from concluded.

*

During the early days of my court case, I had been allowed to keep working at Ngamba Island, but this situation would change. Debby Cox had left the sanctuary by this time and had moved on to other chimpanzee conservation projects beyond Uganda. Debby built such a great team. She surrounded herself with people who had good hearts as well as good skills and intelligence. This was why projects like Ngamba Island became so successful, even though we faced so many big obstacles. But, after Debby left, I believe things changed. There was more focus on people with university education. There was more focus on head knowledge and less on the heart. I think that was a mistake.

At one point after my court trial had started, I was asked to attend a sanctuary managers' meeting in Entebbe. That meeting went on for a very long time—many hours. At the end of the meeting, a large bundle of court documents was produced, and I was asked to explain what these documents were about. I was very surprised.

I asked, "How did you get these papers?"

No one would explain this to me.

Later, when I went to my computer, I found an email that contained all the papers from the court case. I don't know how anyone else could have got this information. It was supposed to be confidential.

By soliciting donations of money for a project other than the Ngamba Island Chimpanzee Sanctuary—that being the Kawuku orphan project—I had breached organizational policy. There was a clause in my employment contract that prohibited me from doing any other fundraising. The court documents became evidence to show that I had breached this contract. I knew then I'd made a mistake. I understood that the consequences of that mistake could have been serious for the Ngamba Island Chimp Sanctuary. The ramifications for me would be devastating.

I think I was naïve and careless. I know I made a mistake. We all make mistakes. However, I know in my heart that I didn't act with any kind of sinister intent. My heart was genuinely for the orphans of Kawuku. And I never ever intended for the chimps at Ngamba Island to be denied any support. Remember, I had refused to take a single piece of fruit away from the chimps, even though my own family was eating dried animal food and cassava paste. I loved those chimps like they were family. I would never take a schilling from them.

The director of Ngamba Island issued me a letter. I was suspended from duty and prohibited from even attending the sanctuary. My income was frozen and the letter promised there would be a full investigation into my conduct.

I remember asking, "When will I be able to return to work?"

No one would answer. They all looked away.

I knew then that there was no hope.

After eight months on suspension, without being able to earn any money, I was permanently dismissed as a caregiver on Ngamba Island. It was over. My beloved job was gone. I remember going home that day. I was very upset. I couldn't eat. I couldn't sleep. But I didn't say anything to Nowera, because I knew if I told her, it would break her heart.

It seemed my time with the chimpanzees of Ngamba Island had come to an end.

Almost.

Although I was dismissed as a member of staff, I was to make a final visit to Ngamba Island. It wasn't to be a happy visit. I was asked to come back for the most somber reason.

Mika, who had been the alpha male of the chimp community on Ngamba Island for about ten years, had gone missing. Sanctuary staff had searched the island, but they could find no trace of him. Eventually, the management began to seek help from others. I suppose they remembered that I knew Mika better than anyone else. And they knew I had found missing chimps on Ngamba Island in the past. So, the sanctuary management contacted me, and they asked if I would go back to help.

I had been deeply disappointed by being fired from Ngamba Island. Still, despite this, I did agree to go back. I wanted to forgive and forget. And I wanted to look for my friend, Mika.

I was asked to lead one of the groups that were searching the island for Mika. It had been more than ten days since he had gone missing. The others had searched and found nothing. Ngamba Island is not so big, but the forest is very thick in places and if you didn't know how chimps behave, finding one individual could be difficult. I was very worried about my friend.

I had known Mika since he was only five years old. When he was little, back at Entebbe Zoo, I remember how he used to ride with me on my bicycle. I remembered giving him his morning porridge and grooming him. He was a beautiful boy. I really loved him.

The first day of the search found nothing. We searched from daybreak until sunset. That night, people were talking, and some of them said they were sure Mika had been killed by a crocodile. There were big Nile Crocodiles in Lake Victoria, and a large one would certainly be capable of ambushing and killing a chimpanzee—even a big male. It would be one explanation for Mika's disappearance, but I wasn't so sure.

The next day, I decided that we should search closer to the sanctuary's electric fences. I thought that maybe, if Mika was sick, he would have come towards the administration area looking for help from the caregivers. It was a good plan. I found Mika's body close to the fence. He was dead. And he had been dead for a long time, because I could see some of his bones. And all around the dead body, I could see where the ground had been trampled and the bushes had been stripped of their leaves and branches. This said to me that the other chimps had been visiting Mika, because he was the boss. I think some of the other males had been displaying around his body.

Chimps understand death. I have seen groups of chimps taking the time to check out a dead comrade. Again and again the group will return, to check, to touch and smell the body. Sometimes you can see they are confused. "Why don't you move?" they seem to be saying. They don't cry, but they may react emotionally to a death. You can see they understand.

I was glad when we found Mika's body, but I was also very disappointed. I felt like maybe there was a chance he could have been found while he was still alive. I remember back when Debby was on the island, we used to have a process where, after two days without being seen, we would start to look for missing chimps in the forest.

There was one chimp—her name was Peace—who went missing when I was working on Ngamba Island. After she had been missing for two days, we started to search for her. I found her lying in a nest, down low in a tree. She had a swollen jaw, and when we took her back to the vet we found she had infected teeth. After the vet gave her medicine, Peace made a full recovery. She still lives on Ngamba Island today.

But you know, Peace was female, and she was a beautiful and mostly gentle chimp. Mika was a full-grown adult male, and he was the alpha. I think this made a difference. I know that some of the staff were afraid of Mika. I used to see this when they were around him. He was a big strong chimp, and he was the boss and he used to scare some of the people. But for me . . . I was always friends with Mika. We knew each other. We trusted each other. So, I worry sometimes that maybe, if I had still been working at Ngamba Island, I could have found him while he was still alive. Maybe the vets could have treated him for whatever was wrong, like they had done with Peace. Maybe he would still be alive today. I do think about this sometimes. It makes me sad. Mika was only twenty-five years old when he died. He was a chimpanzee still very much in his prime.

Ngamba Island gave me a small payment for my work to help find Mika. However, because my bank account had been empty for some time, this amount was swallowed by fees. My financial situation was, once again, very bad. I had no job to go to. I had no way to pay for anything.

Despite my problems, I was determined to become self-sufficient. I traveled around Uganda and bought myself a flock of goats. My goal was to raise the goats on our small plot in Kawuku, eventually selling them for a profit. But from the start, there were problems. Someone had to stay with the goats at all times to prevent them from being stolen. Robberies were very common in Kawuku. Then, some of the goats became sick and therefore couldn't be sold. So, I also tried to buy and sell chickens. I wanted to do whatever I could to help my family. It was not an easy time.

The few animals I did manage to sell failed to raise enough money to even cover my legal costs. As a result, Nowera and I started to sell what few possessions we owned. We owned an old rusting refrigerator, and so we sold that. We sold some other things as well. It felt like we were going back to a life of dust and straw poverty—again.

I wondered how this was justice?

Jim Kurtz—the American man who drank tea with me during Eddie's escape on Ngamba Island, and who had traveled with me across the US—had been in constant contact during these difficult times. Along with the ongoing financial support of Debby, Jim's generosity would save my family. He sent me some money. Oh, God blessed me when Jim shared his heart. I used this money to pay off my bills. I can tell you, Jim was the start of my new life.

Sometimes I wonder about fairness and justice. To have achieved so many good things and to have nothing to show for it, well, it seemed very unfair sometimes. But I had to keep my focus on God. I had to do His will.

After the court case was finished and I was found innocent, some people told me to take revenge on those who had charged me. The pastor from Kawuku was given a bad time by some people. But, you know, I didn't agree with people doing this. For me, forgiveness was the only way. If we don't forgive, then it is our own hearts that remain dirty and wounded. Nobody wins. So, I forgave the people who did these things. I forgave all of them. I believe God wanted me to do this.

Also, I am not bitter towards anyone from Ngamba Island management, the board, or any of the staff who work there. I'm not happy about what happened, and I wish things could have worked out differently, but I accept that I made a mistake. My love for the Ngamba Island chimps has never changed. I want people to still visit and support Ngamba Island, especially to help the chimps there. You know, chimps live for a very long time, and so they need support for many years. Plus, the island is a wonderful place to visit. So, I would say, go to Uganda, because it is a very beautiful country. And make sure you visit Ngamba Island, because you will love it there.

Meanwhile, my life, and that of my family, was in limbo. Job opportunities in Uganda were limited. And, even when I looked beyond Uganda, there wasn't much to offer.

The problem was my education. Despite my experience and expertise, I didn't have any formal college or other academic qualifications. I didn't have a certificate hanging on the wall of my house. I had worked with some of the world's leading chimp experts and I had helped them to do their research. I had worked with people like Debby Cox to perfect new methods of chimp husbandry and management, but I had no paperwork to prove it. So, when jobs were advertised and people with university qualifications applied, I was overlooked.

It was during this time, starting in about 2008, that plans to write my story first began. The plans were supported by Doctor Jane Goodall, who agreed to write the foreword for any book. However, like so many other plans in my life, these plans would take many years to come true. I am very grateful to people like Jim Kurtz, who believed in me and kept working away to have my story told.

And there were some other plans underway. Thanks to dedicated people in Uganda, the United States, and other places, there would be another new opportunity. Wild chimpanzees were disappearing and there were opportunities for someone with experience to work to help prevent those losses. It would mean another new start. It would be another mountain to climb—perhaps the largest mountain I had ever faced.

Once again, it was time for me to move on.

Chapter 11

MOVING ON

Losing my job at Ngamba Island was, and is still, one of the biggest disappointments of my life. And it felt like my life was developing a familiar pattern.

Jim Kurtz put it best when he said, "You know, it's like Stany would work and work to climb up life's mountain, and then someone would come along and knock him back down again."

Jim was right. It did feel like that. And it happened again and again. But I was somebody who had a faith. I was someone who believed in God. I had to trust that He had a plan for me, and that His plan would be worked out in His good time.

His next plans were to be some of the most important jobs I have ever done—and some of the toughest. I was going to have a chance to work on the front line. I was going to try to save wild chimpanzees from persecution and death.

Their scientific name is *Pan troglodytes*. It's a strange name, especially when you think about what those words mean—*Pan* is the Greek god of nature and *Troglodytes* is Greek/Latin for the cave dweller.

So, the name means the "god of nature who lives in caves." I know there are some chimps living in West Africa that do sleep in caves, so maybe that is where the name came from. To me, the common name, chimpanzee, makes much more sense. It comes from one of the Bantu languages and means ape.

Chimps are the most adaptable species of ape, besides human beings. They are found across tropical Africa, from Uganda and Tanzania in the east to Gabon, Liberia, and the Atlantic Ocean in the west. They can live in a wide range of habitats, including rain forest, monsoon woodland, and savannah. But their ability to adapt hasn't been enough to save them from human activity. Chimps are listed as endangered on the IUCN Red List of Threatened Species.[1] And their numbers are rapidly decreasing.

I first learned about the perils facing wild chimps even before I started to work for JGI in Burundi. I would listen to stories from people like Elie, and I became interested in the work of Doctor Jane Goodall. Then, like so many other people, I became inspired by Doctor Jane. I went to work for JGI. And Doctor Jane would become my mentor and my supporter and my friend.

During the years of my childhood, while I was playing barefoot soccer in Burundi with balls made out of banana leaves, Jane Goodall was well into the second decade of her chimpanzee studies at Gombe in Tanzania. Through films and Jane's own books, people had come to learn about her work as well as the lives of the Gombe chimps. Chimps like Flo, Figan, Fifi, David Greybeard, and Goliath became the most famous wild chimpanzees on earth. They also became ambassadors for their kind.

In the early nineties, not long after I'd started working for JGI in Burundi, I had the chance to travel to Gombe myself. This was to be my first real experience of chimps in the wild. It would open

---

1 iucnredlist.org/species/15933/129038584.

my eyes to the work that Doctor Jane had done, and it also helped me to understand more about the differences—and the similarities—between the chimps we had at the halfway house and those living in the wild. And I came to understand how serious the dangers facing wild chimps across Africa were.

When I went to Gombe, I traveled by car through Burundi, down to the Tanzanian border. This was before the civil war had become too bad. At the border we were stopped by Tutsi soldiers. They asked us all kinds of questions, but eventually they let us go across the border. Once we arrived in Tanzania, we caught a boat that sailed down Lake Tanganyika to Gombe, which is on the shore of that giant lake.

That visit would really open my eyes to how amazing Doctor Jane was. It really helped me develop my love for chimps too—seeing them wild and free like that. We arrived at Doctor Jane's camp late at night, so we settled down to sleep and woke early the next morning.

I will never forget that first day at Gombe. We started very early in the morning, before the sun was up, because the guides want to find the chimps while they were still in their nests. At night, chimpanzees sleep safely in nests up in the trees. A nest is a bulky platform of folded and interwoven leafy branches which the chimps make each evening. The nests make a strong and comfortable bed for the chimps and means they can sleep safely in the trees away from leopards and other predators. In the dark green tangle of the rain forest, where sometimes you can only see a few feet ahead, even the largest creature is able to easily vanish from sight, so researchers use nest sites as a way of tracking the movements of the chimps.

After we woke, we had to start climbing up and down hills in the half-light so we could get to the nests from the previous night. It was tough! I remember thinking how much I admired Doctor Jane for doing this same climb over so many years. I was a young, fit guy, and I'd grown up in the African bush. Jane was a girl from England. She must have been really tough! And, you know, at the beginning of her studies, there were no other people out in the forest with her. Her mother came to live with her in camp for a while, but Jane did all the work in the forest by herself.

I remember that first morning, after we found the chimps. It was amazing to watch them wake up. It looked the same as the way the halfway house chimps woke up. When you watch a chimp first wake, you start to see how much like people they are. Some would yawn and stretch, some stood up and peed off the edge of their nests, and some rolled over and closed their eyes and tried to have a sleep in.

And, just like the chimps at the halfway house, all the Gombe individuals had names and personalities. We had heard Doctor Jane tell stories about some of those chimps, and now we were meeting them. There was one name we all knew very well.

One of the big males at Gombe during this time was named Frodo. And everyone knew about Frodo. Son of Fifi—a female chimp Doctor Jane first met at Gombe in the early sixties—Frodo had a very bad reputation. Unlike his mother and his big brother, Frodo was an aggressive bully. He wasn't afraid of humans. He used to throw things. When male chimps display, they charge with bristling hair, sometimes running upright on their back legs, slapping tree trunks, dragging branches, or throwing objects like stones. They will also hit any other chimps who happen to be in the path of their displays. The aim of the display is to show strength and to intimidate rivals. Now, most wild chimpanzees, even the big males, are wary or afraid of human beings. However, at places like Gombe, the chimps had been habituated to the presence of humans. Yet even there, it was rare for a displaying chimp to go near a human. Frodo was different. He would hit anyone. He even used to hit Doctor Jane.

So, the guides said to us, "If we see Frodo coming, we will tell you and you must immediately go to a large tree and wrap your arms around that tree and wait for him to go away."

So, when Frodo did come towards us, the guides called out a warning and we all found ourselves a tree to hang on to. And do you know, Frodo went from tree to tree, and he hit every single one of us on the back. It didn't hurt very much, but he made very sure that he hit all of us. He always acted like he was the big boss, even before he was.

I spent long hours with the Gombe guides, following the chimps as they went about their business. It was fascinating to see how much information the guides and researchers recorded. I couldn't believe that they recorded every single detail. You know, every behavior was important to them. Then, at the end of each day, all the notes were given to a researcher who would type the information into a computer. This was an important lesson for me. It taught me how important it is to watch and understand everything a chimp does. I also learned how the JGI people came to know so much about chimps after I saw them do this work.

While I was at Gombe, I learned a lot more about how wild chimps live. I had the chance to watch their behavior and compare it with the way the chimps at the halfway house behaved. And I also learned a lot about chimpanzee conservation.

My visit to Gombe was probably the first time I really was able to see the kind of threats facing wild chimps. Even at somewhere like Gombe—a place the whole world knew about—there was evidence of human encroachment. And when I thought about what was happening just over the other side of Lake Tanganyika in Congo, I began to understand why so many chimps were being made orphans and brought to sanctuaries like the halfway house.

Learning these things became very important to me. They helped me to do a better job while I was working at the different chimp sanctuaries, but the information also helped me when the time came to move on to the next stage of my career—after I was sacked from Ngamba Island.

Even though I was worried about my own circumstances—being unemployed and not being able to properly care for my family—it was the plight of Africa's chimps that really worried me. It was an important issue for the team at JGI Uganda as well.

The team was finding out about more conservation challenges in Uganda. In a growing number of localities—especially in the west of the country, towards the Congolese border—there were increasing instances of conflict between humans and chimps. These western districts were seeing some of the fastest population growth in the country. With more and more people came a need for more and more land. Agriculture is Uganda's largest industry and accounts for most of the jobs in the country. As agriculture expanded across the fertile land of western Uganda, the forests of the region were being felled and burned. As the forest areas became smaller, chimps and other wild animals lost their source of food and shelter. Animals began to look for other sources of nutrition beyond the forest. And for smart animals like chimps, it didn't take long to discover where the best sources of food were.

Chimps had started raiding crops and farms and even villages in a number of districts. Local people began losing their crops and therefore losing their incomes. Chimps started to be killed. Farmers would lay out steel-jaw traps or wire snares to catch the chimps. Caught in these cruel devices, chimps would often die slow and painful deaths. Sometimes a chimp would be strong enough and smart enough to escape a trap. And in forests across Uganda, you can still find chimps with missing hands or feet or fingers after being injured by traps or snares.

JGI Uganda wanted experienced education officers to travel to these western districts of the country to inform and support the local people. They wanted to explore new ways to help local people earn a living while coexisting with chimps and other forest creatures. Indeed, the Jane Goodall Institute had first been created for just this reason.

Back in the seventies and eighties, as part of her campaign to help wild chimpanzees, Doctor Jane began spending less and less time at Gombe and more time working with local people—especially young people—and their communities. There was one time, while she was flying over the Gombe National Park, that Jane noticed how the chimp forest had become a tiny island surrounded by human activity. The local people were struggling to survive. Jane knew she had to help these local people if there was any hope for saving the Gombe chimps. The people who lived around Gombe were given access to good education, good medical help, and other benefits. They knew that these things have come to their communities because of JGI and that JGI were in the nearby forest because of the chimps. So, the chimps at Gombe were protected by the local people.

Education and advocacy on behalf of local people would become an important component of work undertaken by various Jane Goodall Institutes around the world. Recognizing how important young people were to the future of conservation in 1991, Doctor Jane Goodall created a youth environmental education and advocacy program called Roots & Shoots. From small beginnings in Tanzania, Roots & Shoots is now in around a hundred countries, and the program helps young people to be better informed about conservation and the health of our planet. As Roots & Shoots expanded into more countries, JGI began looking for projects to be developed under the program, including those targeting human-wildlife conflict.

In Uganda, JGI wanted to employ local education officers for the western districts, where human-chimp conflict was getting worse. For me, this was a good fit. I had been very happy helping sanctuary chimpanzees in Burundi and Sweetwaters and Ngamba Island, but at the same time, I always felt sad about what people were doing to chimps in the wild. The conflict and killing weren't just happening in faraway places like the Congo. These things were happening close to where I lived. If someone like me could help the situation, then that was a good thing.

However, there was a problem. Everyone understood there was a need, and people agreed that I would be a good choice for an education officer, but employing me would mean developing a position from nothing, and that was going to cost money. It wasn't just my wages, there needed to be vehicles and other overheads that would have to be paid. We all thought that it was a great idea for a project, but there wasn't any money to make it happen. This, once again, was a situation where my friend Jim Kurtz shared his heart.

Jim began working with JGI as well as a fundraising website called LoveAnimals.org, and he started a crowdfunding campaign. The campaign would eventually raise $16,000 US, which was enough to allow me to be employed as an education assistant for JGI Uganda. Jim arranged other fundraising campaigns as well. These were for Ngamba Island and held during my visits to the United States. As well as raising money for the chimps, some wonderful people in America then created a fund for me, and they raised enough money to allow me to connect electricity to my family house in Kawuku. There was also enough money to install glass into the window frames of the house to help keep the mosquitoes out. I felt so blessed to have this support from people as far away as America. God bless everyone who helped.

With funding now available for one year, in January 2017 I began my role as a JGI Uganda education assistant, working under the Roots & Shoots program. My work area was in a district called Hoima, which was about 140 miles northwest of Kampala, near Lake Albert and the Congolese border.

The Hoima District covered about 1,400 square miles, and it was once covered in lush tropical forest. As the population of the district grew, larger and larger areas of this forest had been cleared for farmland. Even today, most people living in Hoima make their living from agriculture. However, there were also nature reserves in the district as well as small pockets of remnant forest along streams and on hills. As well as in the nature reserves, there were communities of chimps living in some of these smaller forest areas. It was my role to work with residents of thirteen different villages in the district, because many of these villages were close to forests where chimps were still living.

Locals in these villages were coming into conflict with chimps more frequently, as the chimps ran out of food in their small forest homes.

I would travel around the district on a red Honda motorbike. This motorbike was old, and although it had been reconditioned, it wasn't always a reliable mode of transport. Sometimes it would break down, and I would have to find someone to fix it for me. Still, it ran okay most of the time, and I used it to travel around and visit schools and local community groups, where I would provide information about living with chimps.

Walking into the school classrooms, I might see fifty or more students sitting in rows behind their desks. Many of the kids wore school uniforms—bright colored shirts or dresses—but some didn't. I would tell these students stories and give them information about chimps. And I would answer as many of their questions as I could, because I knew these were the adults of tomorrow and they would be the Ugandans making decisions that might destroy or save local wildlife.

I would sometimes go into a local village and speak with large groups of farmers or other local people. Sometimes we would sit in a circle out on the soft grass, under the shade of a big tree, and I would speak to them about chimps and about human-wildlife coexistence. I used to tell them how, with the right knowledge, they could live harmoniously with animals like chimps. Sometimes we would go into a community building and show a film or slideshow. Sometimes we had gifts to hand out, and when this was the case, I would travel in a car rather than on my motorcycle, because there was extra gear to carry.

It was an important job, but it was also very hard. Many of the local people were angry. Residents of these rural villages didn't like the idea of someone coming from Entebbe or Kampala in a fancy car—while being paid a good income—and telling villagers not to kill any more chimps even though the villagers thought the chimps were stealing their livelihoods. I understood how they felt. I knew what it was like to have no money or food to feed my family. So, I had to take the time to understand their point of view. I had to listen to the people and their stories. I couldn't just try and bully them to do what I wanted. I tried to help people understand more about how chimps behave, because I knew it would help these people make better decisions about managing their farms and their livelihoods. But to do that, first I had to invest time with them. The problem was—time was something I didn't have very much of.

With limited funding, my education assistant job lasted for only twelve months. This wasn't long enough for me to make an impact across the whole district. The more I visited each of the villages, the more I realized how much needed to be done. So much forest was still being cut down. The chimps had less and less forest to live in, and that was forcing more of them to break out and go into the villages. But people had been cutting and burning forests in Uganda for generations. This had always been the way they made a living. You cannot hope to change so many years of local customs with just one project in just one year.

Still, even though my time was limited, I was able to help some villages deal with their chimp issues in a way that benefited both the people and the chimps.

In one case we received reports about a group of chimps raiding four villages in the district. The conflict was becoming serious. There had been some clashes, and it was only a matter of time before a chimp or a person was seriously hurt.

The chimps had started stealing crops like bananas, mangoes, and passion fruit while raiding the villages. We heard a story about one man who had seen chimps in a field where he kept some cows. Worried about his cows, the man climbed over the wire fence and went into the field to try and scare the chimps away. He saw an infant chimp sitting close to one of the cows, and when he tried to beat the infant, it began screaming in fear. There were two adult chimps in the field as well, but because there were so many trees and bushes in the field, the man didn't see them. Just like humans, many adult chimps have very strong parental instincts and they will defend an infant ferociously. The sound of the infant in distress infuriated the two adults. They sprang to their feet and began chasing the man. He managed to jump over the barbed wire fence, run to his hut, and lock the door before the chimps caught him. He was very fortunate. If the two chimps had caught him, he could well have been seriously hurt.

When I arrived in each of the villages, I was confronted by angry people. When they found out I was from JGI, their anger only rose. They believed that JGI had come into the villages at night and released the chimps into their fields. They thought we were there to force them off their land, because we liked chimps better than people. It wasn't true, of course, but rumors like this can start easily in Africa, especially in the smaller villages.

In each of the villages, I did some investigations. I asked questions and took photographs. Each village had fields that were close to patches of forest. I found places where the chimps had fed on sugarcane, taken passion fruit from vines, and stripped mangoes from the trees. I even found a chimp's night nest in a tree near one of the villages.

After my investigations, I determined there were only three chimps raiding the villages—two adult females and an infant. I suspected the two adults were the infant's mother and maybe another female relative. I explained to the villagers that the two adult chimps had most probably moved out of the forest and away from their group so that the mother could safely raise her baby during its first months. I know Doctor Jane had recorded this kind of behavior during her studies in Gombe. I told the villagers not to chase or threaten the chimps, just leave them alone. Chimps like to forage, and if they were left alone they would most likely take only a small amount of crops and then they would move on. That was how chimps forage in the forest. If you left them alone, they tended to behave in the same way. I told them the chimps would eventually return to the forest and their own community.

After about six months, we returned to the four villages and found that this had actually happened. The chimps had stopped raiding the crops, and they had moved safely back to their group in the forest. So, in this example, I was able to use behavior education to help people understand the chimps, and the outcome was a good one for both the villagers and the chimps.

Later in 2017, we started receiving reports of a tragedy farther south in the neighboring Kibaale District. There was a family working out in their fields, which were close to small areas of forest where chimps were known to live. The chimps were starving because they had cleared all the food out of their small area of forest. So, they started coming out of the forest to raid gardens and farms. As these raids continued, the chimps grew bolder. On one occasion, some chimps came out of the forest and found a family working in the field. The chimps charged and snatched a six-year-old boy. They dragged the boy into the forest, and there they killed him. When the villagers gave chase, they found the boy lying dead. His stomach had been torn open.

Following this tragedy, JGI considered whether they would send educators to this area to work with the local people, but in the end, they decided we would not go. The people there were very angry and would be unlikely to listen to what we had to say until their anger and grief subsided. My time in the role was running out. Plus, this was a different district from where I had been working, and my funding didn't really cover projects outside Hoima.

Today, as the human population of Uganda's western districts continues to grow, and as the remaining pockets of unprotected forest are cleared, there are more and more incidents of conflict between people and chimps. There is so much need, and there isn't enough money. And too often, the money that is donated doesn't go to the places where it is needed most.

Though I was only able to do the job for one year, the education assistant role was an important part of my career. It gave me a chance to see the issues firsthand, on the ground. That's the only way you can really understand the reality of conservation, and how complex the issues can be. Debby Cox was someone who taught me a lot about these things.

Like Doctor Jane's, Debby's eyes were always looking outward. She knew that to help chimpanzees, you also had to help local people. So, Debby and other JGI staff would go and provide education to the villagers about many different things, not just about chimps. They helped the villagers design and construct charcoal-burning "saving stoves," which were made of clay, retained more heat, and therefore used much less fuel than traditional cooking fires. After these stoves were installed, people living near the forest didn't have to cut down so many trees to get fuel to cook their food.

The people were given seedlings to plant forest trees that bore native fruits. These trees could act as a food source for any chimps leaving their forest home in search of food. They were also given additional crops and trees to replace any that were raided. We used to speak about the benefit of wildlife tourism, like we had done at Ngamba Island and like they do nearby in places like Kibale Forest. Chimps can be worth money to communities. JGI encouraged families to send kids to school, through Roots & Shoots, as we knew that education helped young people understand more about nature as well as more efficient ways of living. Debby and some of the other JGI ladies would even speak to the girls about female hygiene and their periods. The girls were taught to cut and sew their own sanitary pads, which were made of absorbent cloth and could therefore be washed and reused many times.

And, you know, Debby taught me that doing these things helped local people to see that the people who loved the chimps also loved them. We wanted to let these people know we understood and that we wanted to help. I tried to take these lessons from Debby and use them in the year I was the education assistant in Hoima.

*

By the middle of 2018, I was once again without a job. However, 2,500 miles to the south, there had been a series of tragic events at another chimpanzee sanctuary. These events—and the things that followed—meant I was once again called to help out a chimpanzee sanctuary. And this time it was a very important person who asked for my help.

Doctor Jane Goodall wanted to send me to Eden.

Chapter 12

WORKING IN EDEN

In June 2012, a graduate student from the University of Texas was working as a tour guide in a chimpanzee sanctuary near Mbombela (previously known as Nelspruit) in South Africa. According to reports I read from the time, the young man climbed over a four-foot-high safety fence and began walking along the line of an electric fence that was in place to keep the sanctuary chimps from escaping their enclosure. Two adult males named Nikki and Amadeus were close by, on the other side of the electric fence. The young man had made a mistake. He had strayed into the chimps' territory.

At some point, the young man stepped on a rock that was along the fence line, and which the chimps regarded as part of their territory. One of the chimps reached under the fence wire and took hold of the young man by his foot. The young man was dragged under the fence and into the enclosure, and the two chimps began to viciously attack him. With powerful blows and terrifying bites, the attack lasted more than ten minutes. The two chimps seemed determined to kill the young man. If they'd been given the chance, they may well have done so. Fortunately for the young American, sanctuary staff arrived and they began firing shots at the attacking chimps. Nikki was hit, and the shots scared away Amadeus. The staff were then able to reach the student and carry him to safety.

Lucky to be alive, the young man lost both his feet, one of his hands, and received terrible wounds across his face and body. Over the following four years, he would undergo more than twenty separate operations to repair the damage inflicted on his body by the two chimps. He would later be given a bionic hand and two prosthetic feet. I can't imagine what kind of mental and emotional scars that young man must still carry.[1]

Orphans of the bushmeat trade, Nikki and Amadeus were chimps who carried mental scars. They both had troubled pasts. Nikki had been the "pet" of a family who tried to raise him as if he were a human boy. They made him wear clothes and even a wristwatch. His body was completely shaved, except for the hair on his head, which his owners had cut into a little boy style. Even when he grew to be an adult, Nikki continued to have naked arms and legs, because he kept plucking all the hair out. This was possibly a result of the stress from his troubled childhood.

Amadeus too, had suffered a very difficult and cruel upbringing. For many years he had been kept chained outside a petrol station in Angola to attract customers. As he sat there, unable to escape, Amadeus had been taunted and teased continually.

When these two traumatized chimps were first rescued and introduced to each other at Johannesburg Zoo, Nikki and Amadeus bonded as if they were brothers. This relationship continued after they were transferred to a new sanctuary. This sanctuary was run by JGI South Africa and was called Chimp Eden.

After the incident with the American student, Chimp Eden staff took Nikki back to Johannesburg Zoo, where vets removed the bullet he'd received. His wound was not life-threatening, Nikki survived, and

---

1 Shaun Smillie, iol.co.za, November 4, 2017.

once he recovered he was taken back to Chimp Eden. He renewed his friendship with Amadeus.

Two years later, these same two chimps struck again. This time Nikki charged an electric fence while staff were conducting an inspection. He managed to climb both the electric fence and the safety fence, and he attacked one of the staff. Amadeus tried to climb over as well, but he was pushed back by the electric shocks. This time the victim suffered only minor injuries.

The sanctuary decided that both Nikki and Amadeus were too dangerous to be kept any longer. The plan was to euthanize both chimps. This decision was later overturned, and instead, reviews of sanctuary practices were organized.

The incidents at Chimp Eden once again highlighted the dangers of working with habituated chimpanzees. Chimps are naturally excitable, and the adult males can be particularly aggressive—to the point of extreme physical violence. In the wild, there are many recorded instances of male chimps killing other males. Because they have lost their fear of humans, big male chimps in sanctuaries can be especially challenging to work with. In the case of Nikki and Amadeus, because they had suffered terrible abuse at the hands of humans, they carried physical and mental scars, which almost certainly contributed to their violence towards humans. This put caregivers and even sanctuary visitors at risk.

Working with troubled individuals required someone who understood chimp behavior and would take the time to watch, understand, and act on what each of the chimps needed. I had been asked to go to Chimp Eden before, briefly, on two occasions—in 2014 and 2015. I'm afraid those two visits were not very effective or successful. It felt like my input was not respected. I didn't feel like I was able to make much progress.

Now it was 2018. My Roots & Shoots education role had finished, and I was living back in Kawuku with Nowera and my family. It was good to be home, but I wasn't earning any money. Once again, the major thing stopping me from getting a job was my lack of formal qualifications. When I think about those times, it sometimes makes me laugh, because I can also remember being asked where I studied by some very qualified researchers and wildlife professionals.

Doctor Wayne Boardman is a vet who had worked for zoos in both the UK and Australia. He had also contributed his expertise to many conservation projects around the world. In the early days of the Ngamba Island sanctuary, Debby Cox offered Doctor Wayne the chance to visit Uganda and work with the chimps.

"I still remember the first time I anesthetized one of the Ngamba Island chimps," he said a few years later. "We were doing health checks, and I'd only ever used dart guns to sedate chimps before. But Debby came up to me and said, "Sorry, Wayne, but we only hand-inject chimps here." I couldn't believe it. Do you know how close you have to get to a chimp to hand-inject them! My heart was pounding as I walked up to that enclosure, but those chimps were completely connected to Stany. I could see that they respected him and they trusted him. The sedation process was so easy. It was so relaxed. I'd never dealt with chimps like that before. And Stany knew so much about chimp behavior."

It was during these visits to Ngamba Island that Doctor Wayne asked me where I had studied.

"What?" I replied.

"You know . . . university . . . at which university did you do your studies?"

And then I said to him, "No, I didn't go to any university. I learned from people like Debby. I learned by working on the job."

I think that surprised Doctor Wayne. He really thought I had studied at a university, because I was able to communicate so easily with the chimps. And, you know, to have someone of Doctor Wayne's experience speak about me like that . . . well, it made me feel very good. But it didn't get me a job.

Meanwhile, at Chimp Eden, they needed someone to communicate with their chimps. After the troubles they'd had with Nikki and Amadeus, people were asking some serious questions about the sanctuary and the way the chimps were being managed. Perhaps the sanctuary should be closed, but if they did that, then what would be the future for the chimps?

So, JGI South Africa approached Doctor Jane Goodall, and they asked her if she knew of anyone who might be suitable to help Chimp Eden improve the way they managed and cared for their chimps. I was humbled to find out that it was Doctor Jane herself who had recommended me.

After my two short—and unsuccessful—visits to Chimp Eden, I was then offered two longer contracts. The first was in 2018, and the second was a two-year contract that ran from April 2019 to April 2021.

Chimp Eden is located in a popular part of South Africa. It's just off one of the major roads that tourists use when they visit the world-famous Kruger National Park. This has helped Chimp Eden become a very well-visited sanctuary.

When you first arrive at the front gates of the sanctuary, you find tall mud-colored walls and swaying palm trees. It looks more like a scene from a biblical temple. Visitors to the sanctuary can take guided tours, which are held throughout the day. The tours take the visitors to see chimps living in two outdoor enclosures. These two one-acre habitats are separated by fenced corridors, and visitors can watch the chimps from the safety of elevated viewing platforms.

The sanctuary is in the middle of a nature reserve, and the land is a patchwork of grassland and thickets of trees and shrubs. There are administration areas, a kitchen, food preparation rooms, storerooms, vet facilities, and other management buildings. There is also staff accommodation on site.

Because I had to travel from Uganda to South Africa, I was able to live in the on-site accommodation. My home was a small circular hut, with concrete walls and a high-pitched timber-and-thatch roof. Inside, my home was small but very comfortable, with a white-tiled floor, some closet space, and a big, comfortable bed to sleep in. I would wake at about six in the morning and had plenty of time to get myself ready to start work at seven. What a luxury—when there was no travel required! No boats to catch, bikes to ride, or thirty-mile walks to get to my job. Under my contract, I would work for a few months at Chimp Eden, and then I was able to fly home to Uganda to spend a few weeks with Nowera and my family before coming back to South Africa again.

There were around thirty chimps living in the sanctuary during my time there. As was the case in Burundi and Uganda, many of the chimps at Chimp Eden had been stolen from their natural habitats, more victims of bushmeat poaching and the illicit pet trade. Many of the chimps had been smuggled from the Congo to countries like Sudan or Angola before they were confiscated by authorities or rescued by JGI.

The Chimp Eden residents were divided into three separate community groups. Groups one and two had access to the two fenced outdoor enclosures. These two groups also had access to their own fully enclosed night areas for sleeping. Group three included Amadeus—one of the chimps involved in the earlier attacks—and this group did not go out into the open areas, but remained in a secure, fully enclosed home across the other side of the sanctuary.

Just like Nikki and Amadeus, so many of the chimps living at Chimp Eden had suffered terribly in their early lives. Cozy was a member of group two at Chimp Eden. With his spotted pink face, his crazy smile, and his funny yellow teeth, Cozy was a favorite among staff and visitors. He was another one of those cute "performing" chimps—just like the ones I had seen on the streets on Bujumbura when I was a boy. He had originally come from a laboratory in the US, and he then was taken to Italy where he would "pose" for photographs and do tricks for tourists. He had been castrated. He suffered constant abuse and beatings from his owner. These left Cozy with permanent brain damage. When his owner died, Cozy was locked in a tiny cage inside a camper, and he was kept this way for many years. He became very weak, and when he was finally rescued and released, Cozy couldn't even climb. He came to live at Chimp Eden in 2006 and, slowly, Cozy's life became much better.

When I first arrived at Chimp Eden, in 2014 and 2015, I tried to look for ways to improve the management of chimps like Cozy, along with the members of all three chimp groups. During my first two visits, this was difficult. I didn't have much time at the sanctuary, and some of the people in charge weren't interested to hear what I had to say.

Jana Swart is an operations manager at Chimp Eden. She knew how hard it had been to implement change during my first two visits. "I remember it was tough back in those times," she said to me. "There was some resentment and reluctance to change. So, not much was done."

However, in 2018, when I came back to Chimp Eden, I could tell that things were much better. There had been some changes at the sanctuary as well as some new management. I was to be part of those changes. Jana told me that after I came back, things really started to improve.

And because I had longer contracts, it meant that I had more time to learn what was happening and bring in some proper changes. So, just as I had done at the halfway house, Entebbe Zoo, and Ngamba Island, my first job was to spend as much time as I could with each of the chimps. I talked to them and got to know who they were and what they liked. As well as learning more about each chimp, doing this work meant they all got to know more about me as well. This is a very important part of working with a new group of chimps— acceptance into their community. If you are not accepted into their community, then doing things like getting too close to an enclosure fence can be viewed as aggressive behavior by the chimps. I wondered if this was one cause of the aggression that had been displayed by some of the chimps in the past.

I also began working with the Chimp Eden managers and staff, observing practices and considering where improvements might be made. I found there were issues with things like feeding. The keepers were doing their best, but I could see that some things could be done better. At feeding time, the keepers would put the chimps' food out in one pile in one section of the indoor enclosures. This was the process because the staff were worried that feeding the chimps outdoors would make it hard to bring them back inside when it was time. If the chimps knew they would be fed in the night area, they would readily come inside when they were called. So, there was some logic behind this approach to feeding.

However, feeding this way also created problems. It meant the dominant chimps in each group were eating most of the food. I could see that some of the lower-ranking chimps were missing out and not getting enough to eat. I was worried about the physical condition of some of these chimps.

So, I suggested a new feeding regimen, involving enrichment and a more equitable distribution of food among all the chimps in each group. We cut fruit and vegetables into smaller pieces and these were scattered around all three of the enclosures. Doing this meant the chimps had to move all around their enclosure to find their food. They spread out, and this gave each chimp their own personal space as well as an opportunity to find enough to eat. No one chimp—even if they were dominant—could manage to collect all the food for himself, so everyone had the opportunity to get good nutrition.

Also, for the two groups in the outdoor enclosures, scattering their food introduced more natural foraging behaviors. In the wild, chimps need to forage over large areas to find enough food to eat. Searching for food is a big part of their day. At some zoos and sanctuaries, you will see chimps constantly begging for food at the fences. We were able to reduce this habit among the groups at Chimp Eden by scattering their food and forcing them to spend more time foraging.

There were other benefits to this new feeding process. Feeding groups one and two outside meant that visitors could see them for longer. When the chimps were inside their night enclosures, the visitors couldn't see them. And, you know, when a visitor comes to the sanctuary, they pay money to see chimps. When they go up onto the viewing platforms and see chimps behaving in a natural way, this makes the visitors happy. They take photographs and videos of the chimps, and they tell their friends to visit the sanctuary as well.

Another one of my important jobs was to improve the cleanliness of the enclosures—especially the indoor enclosures—and to make all the enclosures more enriching for the chimps. The indoor night enclosures needed some very serious cleaning. I can remember seeing maggots crawling on the floors—which brought back bad memories from my time at the halfway house when we couldn't get down to the sanctuary and the chimp enclosures would be left in a filthy condition. So, one of the first big jobs at Chimp Eden was to suggest that the floors of the night enclosures be resurfaced. Once this was done, we introduced a process where the floors were cleaned every day with disinfectant to keep them clean and help keep the chimps healthy.

I also suggested we build water ponds in the two outdoor enclosures as well as improve water supplies to the night areas. Even though most chimps don't like being in deep water, many enjoy playing with water, especially on hot days. And having cool, fresh water to drink is always a good thing! We installed ropes and old tires for enrichment, as the younger chimps like to swing and the adult males like to bash and hit objects to make noise when they display. We installed hammocks high up in the night enclosures to allow the chimps the feeling of nesting up in the trees like their wild cousins do. And though there were trees in the two outdoor enclosures, these were starting to look very worn and damaged due to the ongoing attention of the chimps, so we had climbing structures built throughout. Being able to climb added another dimension for the chimps. It allowed them to exercise and to enjoy the sensation of being high off the ground. It also gave individuals more options to separate from each other during fights or other stressful situations.

I also introduced the jackpot enrichment technique to Chimp Eden. I had first started using this technique at Ngamba Island, and it gave the staff another way to manage the chimps without the need for sedation or physical separation.

As time went on, it made me very happy to see a change in the chimps. The overall physical condition of the chimps—especially the lower-ranking individuals—started to improve. We began to see less unnatural behavior and even less aggression—and by this I mean unnatural aggression. There should always be a degree of excitement and assertive behavior in a healthy chimp group. It's what happens in the wild. It's the way chimps work out their structure and status and how group dynamics are maintained. But it was good to see much less aggression shown to staff by the chimps. This was a sign, I think, that they were becoming healthier and happier.

I remember Jana saying, "Stany . . . I think you could say he has the 'chimp factor.' I think he can see things in the chimps that others can't. It's like he can feel what they are feeling."

I suppose what she said is true. I do know chimps. I like them. They like me. When I work with them I want to live in community with them, like we are family. And when you belong to a family, you understand each other's needs so much better. Sometimes I might see a change in the skin color of a chimp. Or I might see when they were feeling sad or depressed. For example, I noticed one of the females was having a little bit of difficulty breathing. No one else had noticed it. The vets did an examination of this female and found that she had a bullet fragment lodged in her head, which no one knew about. The vets think this was causing some of this chimp's issues.

Helping chimps in this way has always been a blessing, and so it was at Chimp Eden. It made me feel very good to help them live better lives.

Just as at the halfway house and on Ngamba Island, I began to develop special friendships with some of the chimps at Chimp Eden. And, just as it had been with Mika on Ngamba Island, one of the alpha males at Chimp Eden became my friend. His name was Thomas, and like Mika, Thomas was a fine leader.

Thomas was born in 2004. He had been kept as a pet by an officer in the South Sudanese Ministry of Environment. Eventually, he was handed over to a safe house in the Sudan, because the South Sudanese people were fighting to be an independent country and they were eager to show the world their commitment to stopping the illegal wildlife trade.

When Thomas came to Chimp Eden, he was still an adolescent. He soon grew. And grew. And grew! As he developed, it began to look like Thomas's thin face had been transplanted into a much bigger head. He grew to be a very big boy indeed, with thick black and silver hair and a powerful body. He became the alpha male of group two. But though he was so big and strong, Thomas was mostly a gentle giant. When he was an adolescent, Thomas used to look after other orphans when they became scared or got into fights with other chimps. As alpha, he was a kind and compassionate boss, even sharing his food with lower-ranked chimps, which alpha males often won't do. If another male created trouble in the group, Thomas would be quick to settle things down. I always thought his kindness was amazing, considering the torment of his early life.

And I remember one incredible incident, which just showed what a great alpha Thomas was. One of the lower-ranking males, Charlie, escaped from his enclosure. The staff faced a difficult situation. When Charlie escaped, he liked to cause mischief. He would chase the staff, making them scream and run for safety. Remember, this was a sanctuary where staff had been injured by chimps in the past, and that meant many people were scared of him. I knew something had to be done, and quickly.

However, before I had time to put any plans into action, Thomas came to the rescue. He jumped out of the enclosure as well. But, instead of chasing the staff and causing mischief, he ran straight for Charlie. He ran on his knuckles and his feet, in silent determination. Charlie started screaming. When he caught up with him, Thomas gave the younger male a good beating. Then Thomas forced Charlie back to the fence and into the enclosure. It was wonderful. You should have seen Thomas—he was like a special agent on a mission! He could have joined in the fun with Charlie. He could have found some extra food and some toys to play with. He could have showed some dominance to the humans, but Thomas knew that these were the wrong things to do. He was the boss, and he had responsibilities. He was such a good boy that day! As we already knew, it could be very difficult, dangerous, and expensive for staff to recapture a chimp that escaped, but we didn't need to worry. Thomas did the job for us. It wasn't dangerous at all. It didn't cost us anything. And it only took Thomas a few minutes.

Charlie has since passed away, but Thomas continues to be a strong, compassionate, and effective alpha male of group two. He's a gentle giant. He's my friend.

I am so grateful that Doctor Jane recommended me to come to Chimp Eden. Being able to work closely with chimps again, to make friends with them, and to see them grow to be happy and healthy . . . this is something I am deeply grateful for.

Jana told me the staff were happier at their work as well. And— until the COVID-19 virus shut down tourism in 2020—we were seeing more and more visitors come to the sanctuary, which brought more income for the chimps and for JGI.

"I think we are well on our way to being a self-sufficient chimp sanctuary," Jana said to me. "And I think that could be unique in Africa."

After the terrible feelings I'd had when I left Ngamba Island, to be once again making a real difference to the lives of another group of chimps was a big blessing. My days were once again productive and positive.

And then I was told by some friends and supporters that they wanted to make a documentary film about my life and work. I couldn't believe it.

I have been in some films and on TV shows before. However, these were documentaries about the chimps, not me. I might have been asked one or two questions about the chimps but nothing more. Mostly, I would just be in the background while the cameras were filming someone more important, like the boss of Ngamba Island. To find out there was going to be a film made about me . . . that felt very good. I felt humble.

The filming process was very interesting. They took so much footage! I remember one of the films they made was with a special remote-control camera. They used this camera to get shots of the chimps inside their enclosures. But, while they were using this camera with group two, a chimp named Sally found it. As I have said a few times, chimps are very strong and very curious. If they find something new in their enclosure, they will inspect it. Sometimes they are rough with their inspections. I can tell you, that remote camera was in bad shape after Sally had finished with it. I hope it wasn't too expensive!

I also remember when they filmed Marco, one of the members of group one at Chimp Eden. While they filmed, I greeted Marco with a pant-hoot. This caused him to jump high into the air, doing a backwards somersault. In the documentary, you will see Marco doing this backflip. It was a great demonstration of how athletic chimps are, but it was also evidence of the darker things many sanctuary chimps have experienced. Marco's mother was shot for bushmeat by poachers in the Congo. Marco was only a baby. He was stolen from his dead mother's body and smuggled into the Sudan. The Sudanese military confiscated him, and he was taken to a safe house. Here he was introduced to other infants, and he came to Chimp Eden in 2008. Chimps like Marco often learn new behaviors when they are in unsuitable captive situations. Because they don't have to forage for their food, like their wild cousins, they begin begging for food from their human keepers. Some of them will learn to do different tricks in order to get special treats. So, when Marco did his backflip, it wasn't because he was feeling joyful or because he wanted to be a film star, it was an effort to get a treat from me. Bananas and bread were among Marco's favorite treats.

As I mentioned earlier, one of the things I have done at Chimp Eden is to work with the staff to try and "wash away" unnatural behaviors of the chimps. So, by adding behavioral enrichment to their daily routines—like spreading food out across the enclosures to stimulate foraging rather than begging—we start to reduce the need for the chimps to do these unnatural things. Over time, they start to behave more like normal chimps. So, while Marco's backflips are fun to watch, I hope one day he only does them for fun and not because he wants a treat.

Working at Chimp Eden has meant very good things for me. It has meant good pay and working conditions, but it has also meant being away from home for longer periods than ever before. South Africa is a long way away from Uganda—more than 2,500 miles. I couldn't just drive home for the weekend, and airfare costs a lot of money.

And then the COVID-19 pandemic began. As I speak with you, South Africa has restricted travel with the rest of Africa and the world. I could not return to Kawuku to see Nowera and my kids. There were no flights. For a time, there was no travel at all. So, I had to call my family. I am not so good with technology, and some types of calls are very expensive, but my kids have WhatsApp on their phones, so I was able to communicate with my family using WhatsApp.

Many people have been catching the virus and getting sick, both in South Africa and Uganda. Many people are dying. And, because chimps are so biologically close to humans, they can catch the virus as well. This meant, for many months, no tourists were allowed into Chimp Eden. They say it might be years before international tourism begins again. Only a few local visitors can come. Fewer tourists mean less money for the sanctuary. And less money means the sanctuary has to cut costs. This means cutting staff. In July 2021, I received a new contract to remain working at Chimp Eden, but I don't know how much longer this will continue.

So, the future remains unclear. Despite the successes I have had working in many different sanctuaries and situations, I don't know what might happen to me in the long term. And, after I am finally finished here, I don't know what I will do next. There are no plans for my future. Perhaps I will keep working in South Africa? Perhaps I will work in other parts of Africa with chimpanzees? I worry that my family will have to once again go without things because I'm not earning any money. One day I would like to live and work on our land in Kawuku, and be a farmer, like my father was.

Yet it's important that I don't worry too much about the future. I must remain positive and focused. I believe God will take care of me and my family. He has given me a vision and mission in my life—to love people and to help save chimps. And God has always been faithful. Because of this, I believe that I must continue to live a life of love. I must be there to carry other people. I must have the grace to let others carry me when times are tough.

And sometimes, in Africa, times are very tough—for people and nature. That's why I believe we all have to work together. I believe that God is calling us to ride in the same truck.

Chapter 13

RIDING THE SAME TRUCK

I watch and I see two men walking along a narrow forest path. They're both wearing ragged shorts and sleeveless T-shirts, and they're both barefoot. One of the men carries a very old-looking gun. It is a shotgun, with a barrel like a cannon. The men walk so quietly, it looks like their feet know exactly where they are supposed to tread without making any sound.

At one point, the men pause. Their eyes scan the forest to the left and the right. Their ears know the music of the forest. And they know this is a good place. So, the two men squat down on the moist, leafy ground, and they begin to make the call. They might make the call of a piglet or duiker. The call doesn't matter, so long as they remain patient.

And soon, their prey approaches. In minutes they start to arrive, swaying the branches and screaming loud enough to wake the dead. I watch as the man with the shotgun reaches into the pocket of his shorts and pulls out a red plastic cartridge. He loads the gun and then he fires. It's hard to miss with a shotgun, especially when it's so powerful, and when the prey comes so close. The men smile. As the screams vanish between the trees, I see the two men get to their feet and wade through the undergrowth to find their prize. Soon enough, they will be back in their forest campsite. They will dance and sing, and they will drink *lotoko* to celebrate, for it has been a very good day.

In a clearing hacked out of the undergrowth, I see an old tarpaulin and a smoking fire. Sitting over that fire is a large black pot, full of steaming meaty soup. The meat looks a bit like pork, but I can see the headless corpses nearby, and those bodies have hands. I can count the five fingers. No pig has hands with five fingers.

Close to the headless bodies, lying in a cardboard box alone and afraid, I see a tiny dark-haired infant. He gazes out from between folds of a filthy red blanket; his elfin face is heart-shaped and pale, and his eyes are wide in fear. Everything he knows has been taken away—everything in the world. There are those burning smells and the sounds of the drunken men. And up through the blankets he can see the light shining through the leaves. These were the same treetops he and his mother were climbing through only this morning.

These were scenes from a video on YouTube. Horrible scenes. But they are scenes that are repeated across Africa every single day. And they are scenes that you can't stop by hitting a button or tapping on your screen.

Too often, these are the life experiences of the chimpanzees I have introduced you to throughout this story. These chimps are the victims of human greed and injustice. But sadly, many of the men who hunt those same chimpanzees are also victims in this story. Because, as I have come to learn over many years, the story of conservation in Africa is a very complex one.

Every species of living thing uses the resources of its environment to survive. This is a rule of science and a law of the created order. Human beings are no different. For millions of years, people have used plants and animals for food and fuel and shelter. Those born to the rain forests of Africa used the resources of those forests for their own survival. They understood nature, and they lived as part of it.

But times changed. I read somewhere that Africa has become the fastest urbanizing continent on earth. Even the most remote parts of the Congo basin—the place Europeans once called the "Heart of Darkness"—are being opened to deforestation. People from other continents have discovered Africa's natural resources, and they are prepared to pay money to access them.

And we all have a part to play in the demise of Africa's wild places. Even today, as you read this, you might be carrying a little piece of the Congo in your purse or your pocket. Everyone loves their cell phones and their screens. A mineral called coltan, which is mined in places like the Congo, is used in those devices. As the demand for these minerals grows, even more forest is destroyed. And the more forest is destroyed, the fewer places there are for wild chimps—and gorillas and elephants and all the other forest animals—to live.

When Jane Goodall first started her work at Gombe in the sixties, scientists think there were more than one million chimpanzees living wild across Africa. Today, that number may be as low as 200,000.[1] This means there has been an 80 percent decrease in chimp numbers during my lifetime. There are not many places—even in national parks—where chimps are truly safe anymore. I have seen this with my own eyes. I fear the future for wild chimps is very bleak.

Since I was nineteen years old, I have been working with JGI, and with people like Debby Cox, to help save chimps and to look after the orphans of the forest. I've experienced the impact of human greed and corruption. But I have also come to learn that the issues of conservation are not as simple or as straightforward as many people say they are.

For more than thirty years, I have worked with both African and non-African people. And, you know, the Africans will say to me,

---

1 janegoodall.org.uk/chimpanzees.

"Stany, why do you bother trying to save chimps? Saving chimps or elephants or gorillas . . . that's the white man's business."

I even know of cases where African people have gone and killed wild animals not just for the money but because they want to take revenge against what white people have done to Africans over many centuries.

But then, some white people will come up to me and say, "Stany, tell us how we can stop Africans killing all the beautiful wild animals!"

What these people often don't understand is that many Africans who kill wildlife are themselves living in very bad conditions. Many live in terrible poverty. They face disease and starvation. They can't even afford to feed their own kids. Shooting an elephant for its tusks or a mother chimp for her baby must seem a small price to pay if it means you will be able to feed your own children.

So, for everyone involved in conservation, we have to consider all sides of the story. It's important that the people who can afford to donate to conservation continue to do this. I think it is very important that governments—in all countries and not just Africa—focus on efforts to help wildlife and the people living alongside them.

Conservation projects that don't consider local people will not succeed. I saw this for myself in Hoima. The key to success was working with the local people. That meant traveling to each village and speaking with the people, hearing their stories and accepting their anger and abuse. You cannot do this from a shining office building in New York or Sydney or anywhere else. You must be on the ground. Only then, when local people work together with NGOs and other agencies, and when local people see some actual benefit from conservation, will we have a chance. So, if I could say one thing to people who want to donate to conservation projects, please make sure you donate to projects that help both wildlife and local people.

You know, Debby Cox was someone who taught me a lot about this. Over the years, Debby has raised hundreds of thousands—I think maybe even millions—of dollars for conservation. But Debby never paid herself a large salary. She never owned a flashy car, and Debby never had a big house by the ocean. Even when she owned very little herself, Debby showed such grace and generosity to me and my family. And Debby was never the type of boss who stood back while others did the hard jobs. She would do the same jobs as everyone else did. She was always generous with her time, and her heart was always for local people as well as wildlife. People like Debby taught me we all have a job to do. We all have to ride in the same truck!

At the end of this book, you will find some ideas, tips, and links to websites. These might help you to help chimps as well as the people who share their forest homes. I hope you will have a look and be inspired to help.

Perhaps, as you have been reading through these chimpanzee stories and meeting some of my wonderful friends—Max and Poco and Robbie and Mika and Thomas and Cozy—you might be wondering why these intelligent and beautiful animals are living behind electric fences or wire mesh. Why were they not returned to the forest from where they first came? I know there are people who don't believe there should be any zoos or sanctuaries and the focus of conservation should always be animals in the wild.

Sometimes, people ask me, "Stany, once a chimp is rescued and rehabilitated, why can't you teach it to go back to the forest and live in the wild again?"

I think it's a good question, but the truth is that it can be very hard for chimps to go back to the wild. It can be possible, but it is very hard. If a young chimp is taken from its mom when still a baby, then it won't know how to live in the forest. Living as a wild chimp is very, very complicated. This is why baby chimps stay with their mothers for longer than most other animals. They have to learn what food is good to eat and what will make them sick. They have to learn about using tools to get their food and how to hunt and catch other animals. They must learn about living in a complex community. Chimps who are orphans usually don't know about these things. They have been raised by humans and not chimps, and, while good caregivers do try to "wash away" some of the unhelpful behaviors, it is very difficult to teach a chimp how to live completely wild in the forest, where it will receive no more help from people.

Even though the clearing of forest is happening at a rapid rate, Africa is very large, and there are still big areas of forest that have not been cleared. So, you might think it would be possible for fully rehabilitated chimps to go back to those forests. In theory, it is possible. There are many things to consider, however.

Before you could release any chimps back into the wild, you would have to understand the place you want to send them to. First of all, chimps live in communities, so you can't just take a single chimp and release it and expect it to survive. A group of chimps who know each other and have a hierarchy must be released if they are to have any chance. So, the area of forest you planned to release them must be able to accommodate and feed a large group of animals, and for year after year. Also, if there were wild chimp groups already living in an area of forest, it would be very difficult to reintroduce a new group into that area. Chimps are territorial. Adding new chimps into the territories of experienced chimps who know the land would be like signing the death sentence for those new chimps.

If there were large enough areas of suitable forest without wild chimps living there, we would have to be realistic and ask why. Has disease or hunting or other factors led to the disappearance of the wild chimps? Because, if those factors still existed, then the chimps you released would face a very tough time. Again, they most likely would die.

When we consider these kinds of issues, we start to realize that there are limited opportunities for orphaned chimps to be returned to the wild. So, when chimps are confiscated, what are we to do? I believe good sanctuaries are one of the solutions. Though they are not truly wild, chimps in good facilities receive the best nutrition and vet care and often live longer lives than their wild cousins. The world isn't ideal, but through good zoos and sanctuaries, the plight of wild chimpanzees can be communicated to the wider world. I believe my job, as someone who has worked with chimps for so long, is to help be their voice. I need to tell their stories. I need to fight in their corner. I am riding in their truck.

As you read my story, you might have noticed that I speak a lot about God. I like to read the Bible. I love to go to church. I sing and I pray and I dream. These are the ways God speaks to me. He gives me direction in my life. He has kept me safe in the most dangerous of times.

I believe that God calls us all to love. By this I mean practical love, not just the love you have for your wife or your husband or your kids. The type of love God speaks about in the Bible is love that puts other people first, even if those people are your enemies. God's love means considering all men and women as your brothers and sisters. For me, following God means doing the things that Jesus did.

When people ask me how Nowera and I could put our kids at risk by adopting little Shadia and caring for her mom, or why we adopted Minani's children, even though one of them has HIV, then this is my answer—because I know it's what God wants us to do. We mustn't just speak words of love. If you claim to be a Christian, then you must live lives of love. I believe this.

Now, many of my friends and colleagues are men and women of nature and science. In many cases, I know these friends and colleagues don't believe in God. In fact, some of these people consider believing in God to be silly or illogical. Some of my friends react badly when they find out that I believe in God. I've even had friends and colleagues ridicule and abuse me. But, you know, this is okay. I respect that everyone has a right to believe in God or not. I still respect and love those people. They can still be my friends. And many of them are my friends. I believe that loving God means we must love everyone, even those who hate and mock us.

Debby once spoke about how my life changed after I truly found God. "After Stany left Burundi and then came to Uganda, he was a man carrying a lot of anger. He wanted justice and revenge against those who had wronged him and his family. And remember, like his brother Minani, Stany was heading down a self-destructive path. He was drinking really heavily. But then he found God. He immediately stopped smoking and drinking. He began to look after himself. He became so much happier and healthier."

I know I'm not perfect. I make mistakes. You have read about some of these mistakes in my story. Those mistakes have come at a terrible cost to me and my family. Yet, in all these times, I know God has been with me. I know I am forgiven.

And, if I am forgiven by God, then I know I must forgive others. If you are still reading these words, I would like to say to you—please forgive those who have done you wrong. Please stop hating each other. Please stop killing each other. And please stop killing chimpanzees and destroying nature.

I know this is easy to say. But I also know how hard it can be to do. Like Debby says, I do know how to hate. I used to hate so many people. I wanted revenge. But now, I don't have any room in me for hate. I know that nothing and no one benefits from hate.

When I was in jail and going through court, and when I lost my job at Ngamba Island, it was a very bad time. I would try to speak about forgiveness, but it was a very hard thing for me to do.

Eventually, I was found to be innocent and some people said to me, "Ahh, Stany, now you can get your revenge!" But I didn't want to do it. I decided then that I really had to forgive. Any of the people involved, you know, I really did forgive them. And when your enemies see that you forgive them, then maybe they can begin to change too. Someone has to be the first person to forgive.

In Burundi, for many years there were so many people crying. So many people were dying. The environment had been destroyed. My dad was shot. My dear mother was burned alive. My brother was tortured and killed. I was very angry. I had so much hate. But, when I really thought about it, those people who did all those things, well by now, many of them are probably dead as well. So many people died in that civil war—Tutsi and Hutu. So how does it help for me to hate them? If I do this, I only hurt myself. So, if I do not hate, then all I can do is forgive. No, I don't hate those people. I don't hate Tutsis. I don't hate anyone. Nothing good comes from hate. In Burundi— even today—there is too much hate. No one has truly won the war in Burundi. People speak of forgiveness, but there is not so much of it happening. What Burundi needs—what we all need—is true forgiveness. I believe that kind of forgiveness washes you clean.

Another thing that's interesting to me is how important forgiveness is to chimps. I've seen chimps fighting. I mean, really, really fighting. They might fight until they have given each other serious injuries. And then, very quickly, they will sit down with each other and they will ask each other for forgiveness. They might kiss and hug each other. A very short time after such terrible fighting, the same chimps will be grooming each other. Group harmony is restored. There is true peace. I think that humans can learn a lot from chimps in this way.

So, whether you believe in God or not—perhaps you believe we are descended from ape ancestors like chimps; perhaps you believe we were uniquely created—whatever you believe, forgiveness is something we can all practice. Forgiveness, well . . . to me it's like medicine for the heart.

I remember a cool morning at Chimp Eden. It was August 2019. A friend from Australia was speaking to me on a video call. His face was on my screen, and it felt like he was standing right beside me, even though he was half a world away.

"Stany," he said to me, "after all that has happened to you, how do you stay so positive?"

"There is much to be sad about from my life," I replied. "But, you know, there is much to be happy about too. For me, I have my life and my family and my friends. We still have chimps in the world, and I still have my work with them. I believe I have these things because God's spirit is in me. All my life, people have tried to kill me, but I am still alive. I have worked very closely with chimps for more than thirty years, and I have been charged and attacked many times, but I have never been seriously hurt. I have all my fingers and toes, and I still have two balls! People have tried to destroy me and my reputation and my career, but I still have work and I have my wife and I have my kids and we are all okay. I am positive because I know that God has been with me through all of these things, and He will be with me forever."

I'm not a young man anymore. I'm in my fifties now. I wonder about the future sometimes. I wonder how much longer I will be able to go on working with chimps. Sometimes I think it might be nice to retire and go back to the simple days of my youth—maybe a small plot of land to grow food with Nowera. I don't know what the future will be, but I do know God will be leading my way.

Before my friend vanished off my screen that August morning, he said to me, "Are we riding the same truck, my brother?" And he smiled.

"Yes," I replied, and as I spoke I could feel the warmth of God in my heart. That expression has been so important to me. It means we are all together, in the same place. And it still means a lot.

"We are all riding in the same truck."

HELPING CHIMPS

So, how can those of us living in places other than Equatorial Africa possibly help chimpanzees? Well, Doctor Jane Goodall has an expression: "What you do makes a difference, and you have to decide what kind of difference you want to make."

Debby Cox adds to this by saying, "We have to act, and act now. In some countries, I believe we have less than twenty years to save chimps, or they will become locally extinct."

Twenty years is not a very long time.

While we may not live or work with chimpanzees in Africa, we can all contribute to their wild future by supporting organizations such as the Jane Goodall Institute. As we have seen throughout Stany's story, JGI has a long history of supporting conservation initiatives across Africa. JGI programs and sanctuaries help not only to care for and save chimpanzees themselves but also to assist local people living in areas adjacent to wildlife habitat. This work continues. Today, JGI programs and initiatives include those which provide information and education, agricultural reform, resource management, and micro loans to help local people set up their own small businesses in sustainable ways.

At the end of this section, you will find web links to various Jane Goodall Institutes.

*

Beyond supporting organizations like JGI, there are many other small but meaningful steps we can all take to help secure the future of wild chimpanzees.

For starters, consider if it is always necessary to upgrade to the latest and greatest electronic devices. Would a current cell or smartphone be enough to meet current needs? What if we all used our current devices for a little longer? And, the next time we do need to upgrade our smart device or cell phone, we don't have to just throw the old one away with the trash. The coltan and other minerals in that device can be recycled, thus reducing the need for fresh mining of these elements in places like Congo, which therefore reduced the need to clear forest. Do your research. Ask some questions. Try a Google search for "mobile phone recycling" for more information relevant to your country.

When shopping for a new smartphone or other electronic device, we can also support manufacturers who used recycled materials and commit to sustainable practices that do not destroy the forest and do not engage local people in slave labor. By searching online for "ethical shopping guides," you can find information on which brands to buy and which to avoid—not just electronic goods but a full range of consumer products.

When we travel—in Africa or anywhere else—we can ensure that we don't support the illicit pet trade by having our photograph taken with baby chimps or other wild animals—no matter how cute or appealing they might seem. Never give money to street peddlers or performers using chimps. As Stany was to learn following his young days laughing at the street chimps of Bujumbura, the funny little chimps you see doing tricks or having their photographs taken have almost certainly been forcibly removed from the arms of their murdered mothers, and many have suffered various forms of ongoing abuse.

*

For more than three decades, Stany has worked with and inside sanctuaries and zoos that care for chimpanzees. If we plan to visit a zoo or sanctuary, we can choose to support those organizations that support conservation as well as practice ethical husbandry of the animals in their care. Not all zoos and sanctuaries do these things. It is worth going online to do some research before committing to visit a zoo or sanctuary. Check the websites and look for links to subjects such as *conservation, education,* and *animal welfare.* The sanctuaries and zoos that Stany has worked with—those named throughout this story and listed in the following pages—are all examples of organizations worthy of your support. And there are many more.

We would like to thank you for taking the time to read Stany's story as well as this information. Our hope and prayer is that it inspires you to take action and to do what you can to make a difference.

STANY AND DAVID

For More Information, Visit:

- *Pant Hoot* film: panthootfilm.com/
- Jane Goodall Institute International: janegoodall .org/
- Jane Goodall Institute Australia: janegoodall.org.au/
- Jane Goodall Institute UK: janegoodall.org.uk/
- Jane Goodall Institute South Africa—Chimp Eden: chimpeden.com/
- Ngamba Island Chimpanzee Sanctuary: ngambaisland.org/
- Chimp Haven: chimphaven.org/
- Center for Great Apes: centerforgreatapes.org/

- Disney Animal Kingdom: disneyworld.disney.go.com/en–eu /destinations/animal-kingdom/
- Taronga Zoo: taronga.org.au/
- Sydney Zoo: sydneyzoo.com/
- Chester Zoo: www.chesterzoo.org/
- Max Planck Institute for Evolutionary Anthropology: eva.mpg.de/
- Ethical shopping guide: thegoodshoppingguide.com/

ACKNOWLEDGMENTS

Stany and David would like to thank the following people for their patience, assistance, encouragement, and support during the writing of this story.

To Jim Kurtz, thank you for your unending faith in this project and for your many years of support, both materially and emotionally.

To those who so willingly contributed interviews, information, love, and support—Barbara and Lewis Hollweg, Innocent Ampeire, Dean and Susanne Anderson, Silver Birungi, Cathie Blissett, Doctor Wayne Boardman, Patty Finch-Dewey, Louise Grossfeldt, Doctor Juliane Kaminski, Janet Kellner, Chris Kerr, Joy Kukiriza, Mary Lewis, Pastor David Mukisa, Elie Nkurikiye, Patty Pat, Richard Reens, Renny Severance, Pauline Stuart, Jana Swart, Patrick van Veen, and Aly Wood.

To a pair of remarkable, courageous, and groundbreaking women—Doctor Jane Goodall and Debby Cox—for your love and support; for your wise counsel and steadying hand; and for all you have done to help chimpanzees and the vanishing wild places of our planet. Thanks really do seem inadequate.

Thank you to our agent, Rita Rosenkranz, and to Cal Barksdale from Arcade Publishing for your support of this project.

Finally, we would like to acknowledge all the chimpanzees we've known and loved over the past four decades. Though some of you have left us, we are grateful for what you've taught us and for the joy you brought to our lives. Special thanks go to Max, Poco, Dufatanye, Eddie, Mika, Robbie, Ikuru, Zakayo, Thomas, Cozy, Claude, Snowy, Lulu, Timmy, Moggli, and Lubutu—pant-hoots to you all!